Contents

Chapter 1: About the crisis

Chapter 2: Impact & Consequences

Chapter 3: Action

Introduction

THE CLIMATE EMERGENCY is Volume 357 in the ***ISSUES*** series. The aim of the series is to offer current, diverse information about important issues in our world, from a UK perspective.

ABOUT THE CLIMATE EMERGENCY

In May 2019 the UK Government declared a climate emergency. Environmental activism has never been more vocal and widespread. Greta Thunberg, David Attenborough and Extinction Rebellion have been omnipresent in the news, urging individuals and governments to act, now. In this book we look at the causes and consequences of this global crisis and possible ways to tackle it.

OUR SOURCES

Titles in the ***ISSUES*** series are designed to function as educational resource books, providing a balanced overview of a specific subject.

The information in our books is comprised of facts, articles and opinions from many different sources, including:

- Newspaper reports and opinion pieces
- Website factsheets
- Magazine and journal articles
- Statistics and surveys
- Government reports
- Literature from special interest groups.

A NOTE ON CRITICAL EVALUATION

Because the information reprinted here is from a number of different sources, readers should bear in mind the origin of the text and whether the source is likely to have a particular bias when presenting information (or when conducting their research). It is hoped that, as you read about the many aspects of the issues explored in this book, you will critically evaluate the information presented.

It is important that you decide whether you are being presented with facts or opinions. Does the writer give a biased or unbiased report? If an opinion is being expressed, do you agree with the writer? Is there potential bias to the 'facts' or statistics behind an article?

ASSIGNMENTS

In the back of this book, you will find a selection of assignments designed to help you engage with the articles you have been reading and to explore your own opinions. Some tasks will take longer than others and there is a mixture of design, writing and research-based activities that you can complete alone or in a group.

FURTHER RESEARCH

At the end of each article we have listed its source and a website that you can visit if you would like to conduct your own research. Please remember to critically evaluate any sources that you consult and consider whether the information you are viewing is accurate and unbiased.

Useful websites

www.bas.co.uk
www.climateoutreach.org
www.edie.net
www.europeansocialsurvey.org
www.globalforestwatch.org
www.gov.uk
www.gvi.co.uk
www.metoffice.gov.uk
www.natcen.com
www.news-decoder.com
www.independent.co.uk
www.inews.co.uk
www.ipcc.ch
www.rebellion.earth
www.theccc.org.uk
www.theconversation.com
www.theguardian.com
www.weforum.org
www.worldwildlife.org
www.wri.org

The Climate Emergency

Editor: Tracy Biram

Volume 357

Independence Educational Publishers

First published by Independence Educational Publishers
The Studio, High Green
Great Shelford
Cambridge CB22 5EG
England

Copyright

Photocopy licence

ISBN-13: 978 1 86168 813 2

Printed in Great Britain

Zenith Print Group

What is a 'climate emergency'?

The UK has declared a climate emergency for the national government in Westminster and devolved administrations of Scotland and Wales. But what does it mean?

By James Evison

What is it?

The climate change emergency passed by the UK Government is a motion rather than legislation and does not change the Government's legally binding targets under international accords, such as Paris, or national legislation, such as the Climate Change Act 2008.

This Labour-led motion is a largely symbolic gesture, which makes the UK the first state in the world to make a declaration of an 'environment and climate change emergency'. It was developed following pressure after the Committee on Climate Change's report on net zero, weeks of protests about perceived government inaction on the issue, political meetings with global youth climate figurehead Greta Thunberg, and several television documentaries on the BBC and Netflix highlighting climate change.

In his reasoning for putting the motion onto the floor of the Commons, leader of the opposition Jeremy Corbyn made a series of comments about UK policy and why an emergency should be put in place. These included missing biodiversity targets, an increase in extreme weather events, avoiding more than a 1.5°C rise, water availability, cuts to Natural England budgets, and increasing the targets set out in the Climate Change Act 2008. However, as it is a motion, rather than binding legislation, none of these desired changes will necessarily be put in place as a result of declaring an emergency.

Environment minister Michael Gove did admit during the Commons debate that the Government needed to do more, and on Tuesday produced pledges from the Government to reduce carbon emissions to 'net zero' but without a timeline.

Presenting the motion, Corbyn said: 'The House must declare an environment and climate emergency. We have no time to waste. We are living in a climate crisis that will spiral dangerously out of control unless we take rapid and dramatic action now. This is no longer about a distant future; we are talking about nothing less than the irreversible destruction of the environment within the lifetimes of Members.'

An abstract concept

There is no precise or accepted international definition of a 'climate emergency', but making such a declaration was a key element of the Extinction Rebellion and youth climate protests that have recently occurred across London, the UK and worldwide, and was a byword for taking immediate action and developing policy to mitigate climate change beyond current government targets and international agreements.

The Government is actually following UK local authorities on this issue, and several councils across the UK have passed motions for a climate emergency in recent months, including in London and also in Bristol, where it has been tied into plans to make the city carbon neutral by 2030.

Although it is hard to find an origin story of the phrase 'climate emergency', Bristol Green Party councillor Carla Denyer could be the first politician to put forward the idea of declaring 'a climate emergency' in November 2018 after it had become common parlance among environmentalists and climate activists. In the motion to Bristol City Council, which was passed, she stated that she was inspired by the recent IPCC report which warned that humanity has 12 years to take emergency action in order to prevent global warming greater than 1.5°C. Above this temperature, the risks to humanity of floods, droughts, extreme heat and poverty become much greater, impacting on hundreds of millions more people - hence the need to declare 'an emergency'.

Ambition into action

In terms of turning the declaration into action, comments made by former leader of the opposition and Climate Change minister Ed Miliband are telling. This week he said we should be on a 'war footing' when it came to tackling global warming, and referenced programmes such as the nationwide move from town to natural gas in the 1960s and 1970s as an example of the kind of large-scale projects that will be required. Therefore, it is expected that many of the programmes, projects and policies associated with net-zero and science-based target setting will be used to react to the climate emergency.

Perhaps a crucial difference – and why the Government has chosen to use the alarmist word 'emergency' – is the expectation that these projects should be scaled up from pilots and small schemes into large, company or industry-wide programmes to mitigate climate change and limit temperature rise to 1.5°C.

The emergency is also illustrative of a desire to bring our current 2050 targets forwards – potentially even by decades to 2030 – and even to 2025 as Extinction Rebellion and youth climate leader Greta Thunberg have called for, although this is unlikely. But as a case in point of how fast moving the activity has been around the climate change emergency, this morning (Thursday 2 May) the first minister of Scotland Nicola Sturgeon announced a net zero by 2045 target.

Industry reaction

So how can sustainability professionals respond to the climate emergency? The UK business sector has been a leader in the field of net zero and delivering ambitious goals through the Science-Based Targets initiative (SBTi) to try and reduce the planet's warming to less than 2°C.

But recent plans to drive this down to 1.5°C – which could be seen as being in line with the broad goal of the climate emergency and bringing forward targets – have only been taken up by Tesco, Carlsberg, BT and herbal tea firm Pukka Herbs.

The SBTi has put plans in place to help more businesses convert to 1.5°C targets, a move which could be strengthened by the news this week on the climate emergency, the CCC report and the drive to net zero.

It remains to be seen if targets do change at a UK government level – or if the 'emergency' is simply a symbolic political tool that will be used by the opposition to illustrate a lack of government action in response to the climate strikes.

But it is early days, and it is important to note that the Government's official position to the Labour climate emergency motion was to oppose it. It is also worth noting that current government policy, such as the UK's first deep coal mine in decades (supported by Labour), the expansion of Heathrow, fracking, and continued use of natural gas and oil – as well as 'downgrading' standards outlined in the Environment Bill – illustrates we are some way from seeing the 'climate change emergency' become central government policy.

2 May 2019

This article was produced by **edie** – an industry-leading, purpose-driven business media brand that empowers sustainability, energy and environmental professionals of all levels to make business more sustainable through online content and events. The original article can be read in full here: www.edie.net/news/9/Climate-emergency-what-is-it/

Chart of the day: these countries create most of the world's CO_2 emissions

Just two countries, China and the US, are responsible for more than 40% of the world's CO_2 emissions.

By Sean Fleming

With CO_2 levels still on the rise, being able to track the global emissions hotspots is becoming more important than ever. Before the Industrial Revolution, levels of atmospheric CO_2 were around 280 parts per million (ppm). By 2013, that level had breached the 400ppm mark for the first time.

On 3 June 2019 it stood at 414.40ppm.

15 countries are responsible for more than two-thirds of global CO_2 emissions.

Rank	Country	Emissions in 2017 ($MtCO_2$)	% of Global Emissions
#1	China	9,839	27.2%
#2	United States	5,269	14.6%
#3	India	2,467	6.8%
#4	Russia	1,693	4.7%
#5	Japan	1,205	3.3%
#6	Germany	799	2.2%
#7	Iran	672	1.9%
#8	Saudi Arabia	635	1.8%
#9	South Korea	616	1.7%
#10	Canada	573	1.6%
#11	Mexico	490	1.4%
#12	Indonesia	487	1.3%
#13	Brazil	476	1.3%
#14	South Africa	456	1.3%
#15	Turkey	448	1.2%
	Top 15	26,125	72.2%
	Rest of World	**10,028**	**27.7%**

Image source: Visual Capitalist

There are huge disparities between the world's top 15 CO_2 emissions-generating countries. China creates almost double the emissions of second-placed US, which is in turn responsible for more than twice the level of third-placed India.

Collectively, the top 15 generate 72% of CO_2 emissions. The rest of the world's 180 countries produce nearly 28% of the global total – close to the amount China produces on its own.

Country	tCO2/PERSON	POPULATION
Qatar	49.2	2.6m
Trinidad & Tobago	29.7	1.4m
Kuwait	25.2	4.1m
United Arab Emirates	24.7	9.4m

Per capita CO_2 emissions: the top four

Of course, aggregating emissions by country is just one way of assessing the problem and working out how to counter it. The per capita figures tell a different story.

Here, China doesn't even make the top 20. The per capita No. 1 spot goes to Qatar, with Gulf States making up three of the top four. The US is ranked 8th, behind Australia at 7th.

Looking at per capita figures rather than national-level totals could help bring the reality of the climate crisis closer to individuals. For example, a person may feel their decision to use less-polluting forms of transport is pointless in comparison to the colossal Chinese and American CO_2 figures.

But seeing how population size alters the rankings, and where their country appears, may encourage people to see a connection between their actions and the results they can help bring about.

7 June 2019

Climate emergency mired in economic and political upheavals

Climate change is wreaking havoc. But sceptics remain, while economic and political instability make it hard for political leaders to take tough decisions.

By Sue Landau

The world is stuck between a rock and a hard place.

Three years ago, almost every country in the world agreed to work on limiting global warming to avert a permanent, life-threatening shift in our planet's climate. Thus the ground-breaking Paris Climate Accord was born.

It appeared to herald meaningful international cooperation to halt rising emissions of planet-warming greenhouse gases, after decades of stalling and tentative moves.

But many voices are now clamoring that not nearly enough is being done.

Two top-level reports have just been published warning that countries are not yet even on the road towards the Paris goal, and worse, the goal itself is far from adequate.

Besides that double whammy, climate deniers have been elected in two key countries – in the United States two years ago and in Brazil just a month ago. Climate scientists and activists deplore the inertia they confront even in ostensibly climate-friendly countries.

It is hard to fathom why the world is reacting – or rather, not reacting – like this.

It is incomprehensible if seen purely in terms of the emergency – the droughts, heatwaves and storms now threatening livelihoods and health in different parts of the world, the headlines that regularly tell of dramatic air pollution, freak weather, devastating fires and oceans and wildlife dying off.

> 'Without carbon taxes, how can we cut emissions enough?'

So it may help to understand what other factors are at work to better defuse them. For one, the default position has always been business as usual, and this continues, although it leads to climate disaster.

A big part of such inertia is the weight of existing infrastructure. Coal is a prime example of this. To keep a human-friendly climate, we must stop using it. But coal power plants are still being built, and it still provides most of the world's energy, especially in Asia.

Governments are emerging as a weak link in the climate battle, vulnerable to pressure from industrial lobbies and reluctant to pilot the deep changes needed. Industries are often slow to shift gear and supply practical alternatives. So things go on as before.

There is also a misfit in time scales that pushes climate action down the list of priorities. A week is a long time in politics, a salary is supposed to last a month, but changing energy sources takes decades, and geopolitics plays out on a long-term stage.

Another factor may be the deepening of social inequalities, which is breeding disillusion among those left behind. This can boil over into opposition to ecological changes – as it has in France, where an angry grass-roots movement emerged in November protesting against a carbon tax on heating fuel, diesel and gasoline (petrol).

Carbon taxes are seen as a powerful tool to encourage non-polluting behaviour, but political opposition is strong in many parts of the world. However, without them, how to cut emissions enough?

'Big emission cuts are needed.'

Emissions must be slashed even more drastically than previously thought.

The Paris Accord aims to keep warming between 1.5°C and 2°C by the end of this century. So far, humans have pumped out enough greenhouse gases, chiefly carbon dioxide, to hike the world's average temperature by 1°C since the first industrial revolution. So energy, transport, industry and agriculture must be weaned off fossil fuels by 2050 – a truly vast undertaking.

But those at the pinnacle of monitoring climate are now warning that the 2°C limit is too lax and would leave swathes of the population lacking water and food from mid-century.

In October, a report from the UN Intergovernmental Panel on Climate Change (IPCC) said adopting the tougher target of 1.5°C is the only way to avoid that, and the world has about 12 years left to do what it takes.

Just weeks later, the United Nations' *Emissions Gap Report* said countries must rein in emissions three-fold to achieve the 2°C goal and five-fold to achieve the 1.5°C limit. If this is not done, warming will reach about 3°C by the end of the century.

This is the context for the COP24, this year's meeting of the world's climate diplomats, taking place December 2–14 December in Katowice, Poland – a country incidentally wedded to burning coal for electricity. The meeting is tasked with finishing rules for implementing the targets set in Paris in 2015. Activists are hoping for meaningful road maps to guide actions in different sectors.

'Economic and political instability make for a tough environment.'

Against this backdrop, the protests in France against President Emmanuel Macron's carbon tax are almost a technicality.

By far the most serious climate threats come from President Donald Trump's systematic actions to increase emissions in the United States, the world's second biggest polluter, and the damage Brazil's newly elected president Jair Bolsonaro promises to wreak on the Amazon Forest, the world's biggest carbon sink.

Another potential disaster awaits in China, the world's biggest polluter, if the country does not remedy a disconnect between central government's climate promises and the actions of local officials, according to a report from Endcoal.org.

Hundreds of new coal power plants got approval between 2014 and 2016, and now appear to have been rescheduled by local authorities rather than abandoned. They are roughly equivalent to the entire US coal fleet.

But the French case is different because of the line-up of actors as well as the issue. Here, rural, low-paid people who depend on their cars are challenging a government that sought to make an initial small adjustment towards a low-carbon economy.

The protesters say they are not anti-ecology, just anti-taxes, but they want the carbon tax withdrawn. Their movement is being compared with the unrest of May '68. Yet it also recalls President Trump's electoral base and the constituencies that voted for Britain to leave the European Union.

Some economists discussing the protests on French TV suggested another factor may be ripples from the digital revolution and now the unfolding fourth industrial revolution, which heralds massive economic shifts. The benefits are very unevenly shared, and segments of the population are left behind, which drives political resentments.

Just how such economic upheaval could affect the transitions needed to avert climate change is not clear. But economic and political instability make for a tough environment for ecological transformations.

3 December 2018

Brits know climate change is happening but aren't very concerned about it

Most Brits believe that climate change is happening, but few are very worried about it and only a minority feel very responsible to reduce it, according to the latest British Social Attitudes report.

- Minority thinks climate change is mainly or entirely due to human activity but vast majority believes that climate change is at least in part caused by human activity
- Young people most worried about climate change but least likely to save energy
- Most feel moderate level of personal responsibility and lack of optimism about reducing climate change

The findings, from the European Social Survey (ESS), show that 93% of people acknowledge that the world's climate is at least probably changing. Only 39% say they have given climate change 'a great deal' or 'a lot' of thought. 21% say they have given it very little thought or no thought at all.

Younger and more educated people are more convinced that climate change exist; however, even among graduates and 18- to 34-year-olds, only two-thirds (68% and 66%, respectively) are definite that climate change is happening. In contrast, only half (50%) of over–65s and those educated to GCSE level or below think that the world's climate is definitely changing.

A generational divide in attitudes is also reflected in the fact that younger age groups are more likely to be worried about climate change, with 31% of those aged 18 to 34 reporting that they are 'extremely worried' or 'very worried' compared to 24% of those aged 35 to 64 and 19% of those aged 65 and over.

Equally, graduates express a greater concern about climate change, with 35% saying they are either 'extremely worried' or 'very worried', compared to 20% of those with GCSE or lower qualifications.

36% of respondents say that climate change is mainly or entirely caused by human activity. A majority of people (53%) blame human activity and natural causes equally for climate change, with a vast majority (95%) thinking that climate change is at least in part caused by human activity.

This is almost on par with the amount of people who believe that climate change is at least partly due to natural processes (93%). Only 2% claim that climate change is definitely not happening.

Younger people (46%) are more likely than those over the age of 65 (27%) to think climate change is entirely or mainly due to human activity. 48% of graduates also hold this view compared to just 27% of those educated to GCSE level or below.

Looking at personal responsibility to reduce climate change on a scale of 0–10, the overall average score given by respondents is 6.0. Less than 10% give a score of 2 or less and 15% a score of 9 or 10, while the largest segment of people chose an intermediate score, suggesting a moderate personal responsibility to reduce climate change.

When asked how likely it is that limiting one's own energy use would help reduce climate change, the average score is 4.4, illustrating a low degree of confidence in personal efficacy. The youngest age group, rather than doing the most, are the least likely to often (or very often or always) do things to save energy (71% of those aged 18 to 34 vs 77% of those aged 35 to 64 vs 75% of over–65s).

Respondents think that lots of people limiting their energy use will have a stronger effect on climate change reduction than just one person, giving an average score of 5.8 for collective ability compared with 4.4 on the same 0–10 scale for personal ability. Nevertheless, it is worth noting that the gap between personal ability and collective ability to reduce climate change is not significantly wider.

Furthermore, people are rather pessimistic about the likelihood that many people will actually reduce their energy use, giving this possibility an average score of 3.8. Equally, they have little hope that governments in enough countries will take action as indicated by an average score of 4.3 on the same scale.

The findings show significant differences in attitudes towards climate change by party identification; Liberal Democrats (35%) and Labour (29%) supporters display moderate levels of concern, Conservatives (18%) and UKIP supporters (13%) are less worried.

Although the base is small and so should be treated with caution, Green Party supporters are the most concerned (52%).

In addition, climate change worries differ among those who voted leave (17%) and those who voted remain (32%) in the Brexit referendum. This reflects a more fundamental divide over the existence of climate change. 71% of Remain voters think that climate change is definitely happening, in stark contrast to 53% of Leave voters.

10 July 2018

This information is from the chapter 'Climate Change: Social divisions in belief and behaviour' by Phillips, D., Curtice, J., Phillips, M. and Perry, J. (eds.) (2018), as published in *British Social Attitudes: The 35th Report*. First published 2018.
The climate chapter survey data comes from the European Social Survey.

Most detailed picture yet of changing climate launched

New data gives the most detailed picture yet of temperature, rainfall and sea-level rise over next century.

The UK's most comprehensive picture yet of how the climate could change over the next century has been launched today by Environment Secretary Michael Gove.

Using the latest science from the Met Office and around the world, the UK Climate Projections 2018 (UKCP18) illustrate a range of future climate scenarios until 2100 – showing increasing summer temperatures, more extreme weather and rising sea levels are all on the horizon and urgent international action is needed.

To help homes and businesses plan for the future, the results set out a range of possible outcomes over the next century based on different rates of greenhouse gas emissions into the atmosphere. The high emission scenario shows:

- Summer temperatures could be up to 5.4°C hotter by 2070, while winters could be up to 4.2°C warmer
- The chance of a summer as hot as 2018 is around 50% by 2050
- Sea levels in London could rise by up to 1.15 metres by 2100
- Average summer rainfall could decrease by up to 47 per cent by 2070, while there could be up to 35 per cent more precipitation in winter.

Sea levels are projected to rise over the 21st century and beyond under all emission scenarios – meaning we can expect to see an increase in both the frequency and magnitude of extreme water levels around the UK coastline.

The UK already leads the world in tackling climate change – with emissions reduced by more than 40 per cent since 1990. However, these projections show a future we could face without further action.

UKCP18 can now be used as a tool to guide decision-making and boost resilience – whether that's through increasing flood defences, designing new infrastructure or adjusting ways of farming for drier summers.

Speaking today from the Science Museum in London, Environment Secretary Michael Gove said:

'This cutting-edge science opens our eyes to the extent of the challenge we face, and shows us a future we want to avoid.

The UK is already a global leader in tackling climate change, cutting emissions by more than 40% since 1990 – but we must go further.

By having this detailed picture of our changing climate, we can ensure we have the right infrastructure to cope with weather extremes, homes and businesses can adapt, and we can make decisions for the future accordingly.'

Today's projections are the first major update of climate projections in nearly ten years, building on the success of UKCP09 and ensuring the most up-to-date scientific evidence informs decision-making.

With climate change a global challenge, for the first time, UKCP presents international projections, allowing other nations to use this data to gauge future risks for food supply chains, or check rainfall projections for the likelihood of localised flooding.

Defra's Chief Scientific Adviser Ian Boyd said:

'Climate change will affect everybody. UKCP18 is designed to help everybody make better decisions, from those buying a house to people making large investments in infrastructure. It has been produced using state-of-the-art methods.'

Met Office Chief Scientist Stephen Belcher said:

'The new science in UKCP18 enables us to move from looking at the trends associated with climate change, to describing how seasonal weather patterns will change. For example, heatwaves like the one we experienced in the summer of 2018 could be normal for the UK by mid-century.'

While the UK continues to play a leading role in limiting the causes of global warming and halting temperature rises, some changes to the climate are inevitable. Building on the UK Government's long-term plan for adapting to a changing climate, these projections will help businesses, investors, local authorities, industry and individuals plan for a wide range of possible future changes – alongside taking action to reduce the likelihood of the worst-case scenario becoming reality.

Today's announcement also comes as the UK marks the 10th anniversary of its Climate Change Act – the world's first legally binding legislation to tackle climate change. Just last month the Government hosted Green GB Week – a week of action highlighting the economic opportunities from tackling climate change, encouraging communities and businesses to do more.

While these projections highlight the need for further urgent action, since 1990 the UK has cut emissions by more than 40 per cent while growing the economy by more than two-thirds, the best performance on a per-person basis than any other G7 nation.

Claire Perry, Minister for Energy and Clean Growth said:

'These projections from leading UK scientists build on last month's report from climate experts, highlighting the stark reality that we must do more to tackle climate change in order to avoid devastating impacts on our health and prosperity.

We are already leading the world in the fight against climate change but we cannot be complacent. As we look towards crucial global climate talks in Poland next week, it is clear that now, more than ever, is the time for collective and ambitious action to tackle this urgent challenge.'

While it is not possible to give a precise prediction of how weather and climate will change years into the future, UKCP18 provides a range of outcomes that capture the spread of possible future climates, so we can develop and test robust plans.

The projections will be factored into the UK's flood adaptation planning and the Environment Agency's advice to flood and coastal erosion risk management authorities.

Since 2010, government has invested a record £2.6 billion in flood defences, and we are on track to protect 300,000 more homes from flooding by 2021.

Chair of the Environment Agency, Emma Howard Boyd, said:

'The UKCP18 projections are further evidence that we will see more extreme weather in the future – we need to prepare and adapt now, climate change impacts are already being felt with the record books being re-written.

It is not too late to act. Working together – governments, business and communities – we can mitigate the impacts of climate change and adapt to a different future.

The Environment Agency cannot wall up the country, but will be at the forefront – protecting communities, building resilience, and responding to incidents.'

UKCP18 has been developed by the Met Office Hadley Centre, in partnership with Defra, BEIS, the Devolved Administrations and the Environment Agency, and has been extensively peer reviewed by an independent science panel.

People and businesses will be able to use UKCP18 to explore the types and magnitude of climate change projected for the future, while government will use the projections to inform its adaptation and mitigation planning and decision-making.

26 November 2018

www.gov.uk

Climate emergency: moving from rhetoric to reality

By Adam Corner

On 1 May, 2019, the UK became the first country to declare a 'climate emergency', following similar decisions by a spate of local councils. Last week, *The Guardian* updated its style guide to introduce the terms 'climate emergency, crisis or breakdown' instead of 'climate change', with the newspaper's editor giving the reasoning that these words more accurately 'describe the situation we're in'.

This sequence of emergency declarations are part of an astonishing few months for climate change politics and public engagement. With polls showing a record high in the level of concern about climate change, it's clear that the school strikes, Extinction Rebellion, and the BBC's documentary fronted by Sir David Attenborough have left their mark on public opinion. In one survey, half of respondents (54%) agreed that climate change 'threatens our extinction as a species'. In another, the majority of participants (63%) agreed with the statement ' we are facing a climate emergency'.

So has a social tipping point arrived? Will these declarations of climate emergencies usher in a new era of committed climate change citizenship, as well as joined-up, low-carbon thinking from politicians, and more sustainable business practices? To embed the meaning of a climate emergency into society, public dialogue is crucial – shifting the rhetoric of emergency into a lived reality.

Feeling the fear

The apparent success of the stark, uncompromising Extinction Rebellion messaging (extinction; apocalypse; emergency) reignited a long-standing debate between those who argue that scaring people leads to inaction, and those who argue that spinning climate change as an 'opportunity' obscures the nature of the issue: it might make people feel more motivated, but it misrepresents the problem.

The truth, though, has always been more complex, and more interesting. Research shows that generating negative emotions, such as fear, can resonate (the Extinction Rebellion demonstrations have certainly done this), but that without constructive, practical actions to enable people to do something with the fear they feel, the vital emotional energy can dissipate into a sense of helplessness.

Until recently, climate change hasn't always felt like something to fear in the here and now. With climate impacts now really beginning to bite, there is no shortage of visceral human stories. Fear is tangible and real for many people, not something that needs to be constructed through a message. As our own Climate Visuals research bears out, authenticity in climate communication is key – and right now, fear is an authentic emotion for an increasing number of people.

The need for public dialogue

The idea that climate conversations are an essential part of the transition towards sustainability – that 'talking climate' is an important action in itself – is part of the Climate Outreach DNA. But recently, there's been a huge surge of interest in a particular type of public dialogue, citizens' assemblies, as a means of accelerating climate action.

As one of the core demands made by Extinction Rebellion, citizens' assemblies have moved into focus, with politicians, campaigners and civil society groups racing to back the idea of a national conversation on climate change.

Where citizens' assemblies and other large-scale deliberations have taken place in the past (such as the recent Irish Citizens' Assembly), the recommendations for climate and energy policies have been wide-ranging and progressive, outstripping the ambition of the actual policies in place.

So if a climate emergency is to be more than a rhetorical flourish, we need honest and upfront conversation about what life in a 'climate emergency' means. How can we live differently (and better), in a way that meets the need to decarbonise rapidly? The newly announced ESRC Centre for Climate Change and Social Transformations (CAST), of which Climate Outreach is a core partner, has this question as its starting point.

There is no doubt that climate change is an emergency, but how that powerful idea plays out is all to play for. Talking climate is crucial to translate the drame of declaring a climate emergency into a day-to-day

21 May 2019

Hothouse Earth: our planet has been here before – here's what it looked like

***An article from* The Conversation.**

THE CONVERSATION

By Jan Zalasiewicz, Senior Lecturer in Palaeobiology, University of Leicester

Even if carbon emissions are reduced to hold temperature rises at the 2°C guardrail of the Paris Agreement, changes already afoot in the environment such as melting permafrost and forest die-back could accelerate warming well into the future, potentially pushing our planet into what is being called a 'Hothouse Earth' state.

The risk of a hothouse was raised by a recent study, though the authors stressed that it is not inevitable. But what is a Hothouse Earth state, and how will it feel for humans and the rest of nature?

The Earth has been in hothouse (often called 'greenhouse') states before, and there is not one kind of Hothouse Earth, but several. A little like Dante's circles of Hell, they progress into ever-deeper states of heat and changes to the planet's biosphere and climate. The end result is undoubtedly hellish, and even the early stages would be, for humans at least, decidedly uncomfortable.

Paradise lost

The first state last occurred 125,000 years ago, during the previous interglacial phase of the Ice Ages. Atmospheric carbon dioxide levels were like those of pre-industrial times, at about 280 parts per million (ppm), and global temperatures were generally similar to today. Early humans were present, but in small numbers with only local impact on Earth's ecology. We might regard it as an Eden of pristine landscapes and ecosystems.

During this interglacial phase, sea levels rose to some six metres above today's as part of the Greenland and West Antarctic ice sheets melted, perhaps as changing ocean currents carried extra heat to the ice. It's a reminder of how easily sea level can change. Then, animal (including human) and plant communities simply adapted by migrating with the shoreline. Modern civilisation, with its sprawling coastal megacities, would not adapt so easily.

We are already set for the next stage. By burning fossil fuels we have launched atmospheric carbon dioxide levels beyond 400ppm – that's an extra trillion tons of carbon dioxide in the air. With this thicker thermal blanket, the Earth is absorbing more heat, with most of it going into the oceans. We are already noticing the effects, with extreme weather events rising and ecosystems such as coral reefs already suffering drastic change.

The world at 400ppm CO_2

The last time the Earth saw these kinds of carbon dioxide levels was three million years ago, well before *Homo sapiens* appeared, in what is called the 'Mid Piacenzian Warm Period' of the Pliocene Epoch. This was warm – but not yet

truly hot. The Earth still had a lot of polar ice, especially over Antarctica, but ice on Greenland and West Antarctica was much less extensive, and sea levels were some ten metres or more higher. Global mean temperature was perhaps a couple of degrees warmer than at present, with more warming around the poles than at the equator. If carbon dioxide levels now hold steady, this is the kind of Earth we could be heading towards. This period in Earth's history was also Eden-like, with a diversity of life on land and at sea – but getting to that state may be traumatic for crowded humanity as the sea level keeps rising. A true Hothouse Earth emerged when carbon dioxide levels reached something like 800ppm – about double those of today. This was the world of the dinosaurs, 100 million years ago. There was little or no ice on Earth and the polar regions had forests and dinosaurs which were adapted to living half the year in darkness.

The biosphere thrived, though equatorial regions tested the thermal limits of life. Much of the low-lying land had been claimed by the sea, which was now a worldwide warm bath in which animals steered a course to avoid the large, oxygen-depleted regions, a result of the sluggish ocean currents typical of an ice-free world. Even this type of Earth is not so unpleasant, though – once you're there.

But it's the transition that's tricky. Some combination of unrestrained carbon emissions and the natural feedbacks of greenhouse gas released from melting permafrost and forest dieback might set us on such a trajectory in little more than a century. Humanity, in such a world, might crowd on to the remaining land and mourn its drowned cities.

Hell on Earth?

Hothouse Earths can also get hotter during 'hyperthermal' events, typically triggered by sudden, massive carbon dioxide releases from extraordinary volcanic outbursts. The larger of these coincided with the times of the great dyings – mass extinction events like those at the end of the Permian Period 251 million years ago, in which most life perished through extreme heat, suffocation or starvation. This is where true hell on Earth appears.

The ultimate Hothouse Earth has, thankfully, not yet been reached. If it had, we would not be here to discuss it. It is a runaway greenhouse world like that of our sister planet Venus – heated to the point where the oceans are boiled away, with water vapour streaming through a punctured stratosphere, leaving a furnace-like surface devoid of life.

It seems that even burning all of our hydrocarbon fuels will not yet invite such a state, which is some comfort. However, this is surely the eventual state that the Earth will reach in about one billion years, as the sun heats up. Fortunately, this is not our immediate problem.

Hothouse Earth is a journey with many stops and even the next few steps would be a bumpy ride for human civilisation. It's the speed of the change that's crucial. In a transition stretched out over millennia, humans could probably adapt to even a dinosaur-style hothouse. But if it's going to come in centuries, or even in a human lifetime, there'll be trouble ahead.

13 August 2018

www.theconversation.com

Are hurricanes getting stronger – and is climate breakdown to blame?

By Oliver Milman

What is a hurricane, exactly?

A hurricane is a large rotating storm that forms over tropical or subtropical waters in the Atlantic. These low pressure weather systems draw upon warm water and atmospheric moisture to fuel their strength and will gather pace if not slowed by patches of dry air, crosswinds or landfall.

'They are very tall towers of winds that move at the same speed, sometimes 60,000ft tall,' says Jim Kossin, a scientist at the US National Oceanic and Atmospheric Administration. 'If they are unmolested by wind shear or run over land they will continue on their merry way.'

Storms are given names once they have sustained winds of more than 39mph. Hurricanes are gauged by something called the Saffir-Simpson hurricane wind scale, which runs from one to five and measures speed.

Once a storm gets to category three it is classed as a major hurricane, with winds of at least 111mph and enough force to damage homes and snap trees. Category five storms, of at least 157mph, can raze dwellings, cause widespread power outages and result in scores of deaths.

This strongest class of hurricanes includes Hurricane Katrina, which caused the inundation of New Orleans in 2005, and Hurricane Maria, which flattened much of Puerto Rico in 2017.

How do they differ from typhoons and cyclones?

Both hurricanes and typhoons are tropical cyclones – the only difference is the location where they occur. In the Atlantic, the term 'hurricane' is used, while 'typhoon' is used in the Pacific. In the South Pacific and Indian Ocean, the term 'tropical cyclone' is often deployed.

Why is there a hurricane season?

Almost all hurricanes develop once the northern hemisphere approaches summer, with the hurricane season running from 1 June to 30 November. The season peaks between August and October.

This is because wind shear, which can disrupt hurricanes, dies down during summer, while the temperature of the oceans rise and the amount of moisture in the atmosphere increases. These conditions are ideal for spawning hurricanes.

The season isn't strictly defined, however. 'It can start earlier,' says Jennifer Collins, a hurricane expert at the University of South Florida. 'Recall 2016, when Hurricane Alex (a storm that rattled Bermuda) formed in January.'

What has happened with hurricanes in recent years?

It's been a punishing past few years for people living in the path of hurricanes in the US and Caribbean. Last year there were an above-average 15 named storms, including Hurricane Florence, which brewed off the west African coast before barrelling into North Carolina, plunging much of the state into darkness and dumping up to 76cm (30 inches) of rain in places, resulting in flooding that killed dozens of people.

This was followed by Hurricane Michael, the first storm to make landfall in the US as a category five event since 1992. The 160mph storm obliterated the town of Mexico Beach in Florida, caused more than 70 deaths and racked up an estimated $25 billion (£19 billion) in damage.

These disasters came in the wake of the 2017 hurricane season, which caused a record $282 billion in damage. Hurricane Harvey unloaded 33 trillion gallons of water on Texas, the astonishingly strong Hurricane Irma, which reached a top speed of 177mph, ravaged Florida and several thousand people died in Puerto Rico after Hurricane Maria, another category five storm, tore across the island.

The misery in Puerto Rico, in particular, is ongoing, with the US Government strongly criticised by local elected officials for a sluggish and insufficient response to the catastrophe.

Does this mean that hurricanes are getting stronger and more damaging?

While the overall number of hurricanes has remained roughly the same in recent decades, there is evidence they are intensifying more quickly, resulting in a greater number of the most severe category four and five storms.

The proportion of tropical storms that rapidly strengthen into powerful hurricanes has tripled over the past 30 years, according to recent research. A swift increase in pace over a 24-hour period makes hurricanes less predictable, despite improving hurricane forecasting systems, and more likely to cause widespread damage.

The devastation unleashed by recent hurricanes has led to warnings that premiums may rise as insurers face ballooning claims. A record $135 billion was paid out by insurers in North America in 2017, mostly as a result of hurricane damages. 'We have a new normal,' says Ernst Rauch, a senior executive at insurance company Munich Re. 'We must have on our radar the trend of new magnitudes.'

Is climate breakdown to blame?

A range of factors influence the number of hurricanes smashing into land, from localised weather to periodic climatic events such as El Niño. Prior to 2017, the US had experienced a hurricane 'drought' that had stretched back to Hurricane Wilma in 2005.

But there is growing evidence that the warming of the atmosphere and upper ocean, due to human activity such as burning fossil fuels, is making conditions ripe for fiercer, more destructive hurricanes.

'The past few years have been highly unusual, such as Irma staying strong for so long, or the hurricane in Mozambique that dumped so much rain,' says Kossin. 'All of these things are linked to a warming atmosphere. If you warm things up, over time you will get stronger storms.'

Climate breakdown is tinkering with hurricanes in a variety of ways. More moisture in the air means more rain, while storms are intensifying more quickly but often stalling once they hit land, resulting in torrential downpours that cause horrendous flooding.

Rising sea levels are aiding storm surge whipped up by hurricanes – one study found that Hurricane Sandy in 2012 probably wouldn't have inundated lower Manhattan if it occurred a century previously because the sea was a foot lower then. According to the UN's Intergovernmental Panel on Climate Change, the maximum intensity of hurricanes will increase by about 5% this century.

The expanding band of warmth around the planet's tropical midriff also means a larger area for hurricanes to develop, resulting in fierce storms further north than before, such as Florence. In the Pacific, this change means typhoons' focal point is switching from the Philippines towards Japan.

Researchers are currently attempting to ascertain if climatic changes will help bend the path of hurricanes enough that more will charge in the direction of the UK in the future.

'This has implications for places that have historically been unaffected by tropical cyclones,' says Collins, who added these newly hit areas are likely to suffer a significantly higher risk of structural damage than traditional hurricane zones.

'We are already seeing effects of climate change,' says Collins. 'While there is not consensus on the frequency of hurricanes in a warmer world, there is a consensus that the hurricanes are becoming more intense, and hence their impact will be worse.'

Are people adapting to these changes?

The forecasting of hurricanes has become a fine art, with scientists able to predict with sharp accuracy the anticipated path and ferocity of hurricanes. In the US, affected states have refined systems for warnings and evacuations and have a hefty federal agency, Fema, to plough billions of dollars into patching up shattered towns and lives.

But planning is often haphazard, with flooded houses repeatedly rebuilt in the same locations despite the morphing risks posed by the climate crisis. The concreting of Houston's green spaces removed key sponges for Harvey's water, which sloshed into people's homes. Meanwhile, natural buffers to hurricanes, such as mangroves and coral reefs, are being stripped away around the world as a result of coastal development, pollution and warming waters.

There are more people in harm's way, too – in the south-eastern US, for example, coastal populations grew by more

Why do we name hurricanes anyway?

In short, to avoid confusion when there are two or more active hurricanes. In the past, storms were named after saints, with hurricanes only named from 1950 onwards, using the phonetic alphabet.

The names used have diversified since then, with a list drawn up ahead of each hurricane season and ticked off as each named storm develops. Particularly devastating hurricanes cause names to be retired, which is why we won't see a Hurricane Katrina, for example, again.

From 1953, female names were used for hurricanes, prompting an outcry. "You can imagine that women did not want something as destructive as a hurricane to be associated with their sex alone, so with the feminist movement pushing the issue, in 1979 males and female names alternated," Collins explained.

This does appear to make a difference, at least in psychological terms. In 2014, US researchers found that Americans are less afraid of hurricanes with female names. "People imagining a 'female' hurricane were not as willing to seek shelter," said study co-author Sharon Shavitt.

than 50% from 1980 to 2003. Climate change adaptation rules have been scrapped by Donald Trump's administration, making it easier to build critical infrastructure in risky coastal areas.

'Coastal towns and cities are not currently prepared for the changes already occurring and will continue to occur,' says Collins. 'We know that there are areas that are prone to flooding. We need to not rebuild on these areas, and build on higher ground.

'Those who deny scientific findings in favour of magical thinking and other such fallacies will only leave the world a more unstable and dangerous place for future generations to come."

The situation is even starker for poorer Caribbean nations that will increasingly rely upon international help to deal with stronger hurricanes and rising sea levels.

Hurricane Maria devastated the island of Dominica, leaving just 5% of the country's buildings intact. Its prime minister, Roosevelt Skerrit, who lost his own roof in the storm, subsequently told the UN that he had come 'straight from the frontline of the war on climate change'. 'We as a country and as a region did not start this war against nature' a visibly shaken Skerrit says. 'We did not provoke it. The war has come to us.'

What next?

Researchers have been poring over ocean temperatures and other data to ascertain what's in store for the 2019 hurricane season, which starts on 1 June. US officials will unveil their best guess on Thursday in Washington. Meteorologists at Colorado State University have predicted there will be a slightly below average Atlantic season of 13 named storms, five of which will become hurricanes. This prediction rests on the presence of a mild El Nino – a natural climatic event that periodically warms the Pacific Ocean, a process that tends to suppress the development of Atlantic hurricanes.

There is still plenty of uncertainty in these early predictions. 'Early forecasts can be a but sketchy,' says Kossin. 'In general, it looks like it will be around average. But we will have to see.'

20 May 2019

Enough rainforest to fill 30 football pitches destroyed every minute last year

'The world's forests are now in the emergency room,' says expert.

By Peter Stubley

The world lost 12 million hectares of tropical rainforest last year – an area the size of North Korea and the equivalent of 30 football pitches every minute, according to a new report.

'It's death by a thousand cuts,' said Frances Seymour, senior fellow at the US-based World Resources Institute (WRI), which led the research based on an analysis of satellite imagery.

'The health of the planet is at stake and band aid responses are not enough. The world's forests are now in the emergency room.'

The global destruction of tree cover includes around 3.6 million hectares of primary rainforest – older, untouched trees that absorb more carbon and are harder to replace – covering an area the size of Belgium.

Brazil lost the most tropical primary rainforest in 2018, at 1.3 million hectares, followed by the Democratic Republic of Congo with 481,248 hectares.

The Global Forest Watch report suggested that most of Brazil's loss last year was down to 'cutting in the Amazon' by illegal loggers and militias, which threatened the survival of nearby "uncontacted" indigenous tribes.

It came as Brazil's federal police revealed they had uncovered a scheme to illegally harvest timber in the Amazon region. The criminal conspiracy is said to involve state environmental agency officers and forest engineers.

The greatest increase in deforestation compared to 2017 came in Ghana, with a 60 per cent increase mostly blamed on illegal mining and the expansion of cocoa farms.

'Forests are our greatest defence against climate change and biodiversity loss, but deforestation is getting worse,' said John Sauven, executive director of Greenpeace UK.

'Bold action is needed to tackle this global crisis including restoring lost forests. But unless we stop them being destroyed in the first place, we're just chasing our tail.'

However there was some improvement in Indonesia, where government policies on protected forests appeared to have resulted in tree cover losses dropping to their lowest rate since 2003.

Much of the deforestation in that country is blamed on land clearance for oil-palm plantations.

Last year's total loss of 12 million hectares was the fourth-highest since records began in 2001 but was lower than 2016 and 2017 when losses peaked largely due to forest fires.

25 April 2019

Great Barrier Reef hard coral cover close to record lows

Coral bleaching, crown-of-thorns starfish and cyclones reduced coverage across the reef over past five years.

By Adam Morton, Environment Editor

Hard coral cover on the Great Barrier Reef is near record lows in its northern stretch and in decline in the south, surveys by government scientists have found. A report card by the government's Australian Institute of Marine Science says hard coral cover in the northern region above Cooktown is at just 14% – a slight increase on last year but close to the lowest since monitoring began in 1985.

A series of 'disturbances' – coral bleaching linked to rising water temperatures, crown-of-thorns starfish outbreaks and tropical cyclones – have caused hard coral cover to shrink across much of the world heritage landmark over the past five years. Depending on the location, coral coverage is between 10% and 30%

Mike Emslie, the institute's acting head of long-term monitoring, said the report included glimmers of hope: individual reefs, including those on the outer shelf in the Whitsunday Islands, were found to have lively communities and tiny juvenile corals were discovered across the 2,300km reef system. The density of juvenile coral suggested recovery was possible if there were not further disturbances. He said it indicated there was some resilience in the system but added: 'The important thing is the absence of further disturbances. If we have more coral bleaching events all bets are off.'

The northern and central sections of the reef were hit by back-to-back mass bleaching events as ocean heating increased in early 2016 and 2017, killing vast areas of coral. A study led by Terry Hughes, the director of the ARC Centre of Excellence for Coral Reef Studies, found 30% of coral died after the 2016 heatwave alone.

Intergovernmental Panel on Climate Change scientists estimated 99% of corals across the globe are likely to be lost if the climate crisis is not addressed and global heating reaches 2°C.

The Institute of Marine Science report warns the extent of hard coral in the north may be even lower than the 14% estimated due to skewed surveying – the greatest bleaching damage was on inshore reefs and they were under-represented in surveys due to safety concerns. The highest level of coral reef cover recorded in the north was 30% in 1988.

Emslie said the reef's southern section had escaped the worst effects of coral bleaching and cyclones since 2009 but has been affected by a severe outbreak of crown-of-thorns starfish since 2017. Crown-of-thorns feed on coral and spawns so rapidly it is difficult to tackle once it takes hold. Its spread has been linked to nitrogen from fertiliser and pesticides in agricultural run-off.

He said the starfish had a particularly devastating impact in the Swain Reefs national park, more than 100km off the coast between Rockhampton and Mackay. Across the south, hard coral cover is 24%, down from a high of 43% three decades ago.

The central region, from Airlie Beach to north of Cairns, has been significantly damaged by Tropical Cyclone Debbie in 2017 and the crown-of-thorns' southward spread. Hard coral cover fell from 14% to 12% last year. It was 22% just three years ago.

The broadcaster and natural historian David Attenborough made headlines this week when he told a UK parliamentary committee that the change to the Great Barrier Reef was one of the clearest examples of the climate crisis he had witnessed. He said it was extraordinary that people in power in Australia remained in denial about the scale of the problem.

Environment group the Australian Marine Conservation Society said the latest government data showed coral decline was happening on an unparalleled scale, mainly due to the climate crisis. The society's spokeswoman, Shani Tager, said the reef remained a dynamic place that was home to thousands of animals and supported 64,000 tourism jobs but was in serious trouble.

'We need our governments to act fast,' she said.

The science record card coincided with the release of an annual work plan by the Great Barrier Reef Foundation, a formerly small organisation that was last year granted $443 million in public funding despite not having applied for it.

The foundation plans to spend $58 million this financial year focusing on improving water quality, managing crown-of-thorns starfish and collaborating with the tourism industry to engage visitors in citizen science activities. Anna Marsden, the foundation's managing director, said innovation would be at the heart of a reef restoration and adaptation programme run in collaboration with leading marine science institutions.

Richard Leck, from the World Wide Fund for Nature, said the report card showed Australia needed to urgently reduce its dependence on fossil fuels and accelerate a transition to a clean economy.

11 July 2019

www.theguardian.com

Six animal species and how they are affected by climate change

By Marike Lauwrens

Climate change has become an everyday term as more people become aware of it. But have you thought about what this phenomenon means in reality? Changes in temperature might not seem extensive, but we are already seeing dramatic results in many areas:

- Some islands no longer exist, because of the sea levels rising.
- The occurrence of natural disasters is increasing.
- A number of stunning destinations are on the brink of vanishing.
- Wildlife species are needing increasing protection due to changing ecosystems and habitat loss.

In practice, climate change affects animal species in the following ways:

- Climate patterns change and animals have to adapt accordingly.
- Animals experience habitat loss due to increased greenhouse emissions.
- Animals have to alter their breeding and feeding patterns in order to survive.

If these animal species can't migrate to more favourable climatic areas, their fate might be sealed. Learn more about six animal species, and how they are affected by climate change.

1) Cheetahs

The African cheetah is the world's fastest animal but is racing against its against near-threatened status in the face of climate change. In some areas, the cheetahs' prey populations are declining, and as a result, the cheetahs have had to change their diets.

A rise in temperatures has affected this big cat's ability to reproduce. Male cheetahs have shown lowered testosterone levels and a sperm count almost ten times lower than your house cat.

A rapid decline in free-ranging cheetahs means resources are being used to study and preserve these master survivors in managed parks. For example, GVI volunteers gather data in Karongwe.

2) Giant panda bears

This iconic bear and World Wide Fund for Nature (WWF) trademark feeds exclusively on bamboo, but unfortunately, climate change is causing a major wipe-out of bamboo in their natural habitat in China. Apart from being the bears' staple diet, bamboo also provides them with shelter from the elements.

3) Green turtles

Green turtles are very sensitive to changes in temperature.

A baby turtle's gender depends on the temperature of the sand where the eggs are laid. The warmer areas produce female turtles. With climate change causing an increase in temperatures, more females than males will hatch.

You can support the protection of this endangered species by joining GVI's endangered turtle conservation and research programme in Thailand, or you can travel to Costa Rica to be part of conserving hawksbills, leatherbacks and green sea turtles.

4) Asian elephants

These gentle giants are particularly sensitive to high temperatures. In order to survive, they need to drink a great amount of fresh water daily.

Climate changes and global warming makes it more difficult for elephants to get the water they need. Warmer temperatures also create favourable conditions for invasive plants to thrive and outmatch the elephants' regular food sources.

5) Polar bears

Climate change and global warming result in less Arctic sea ice for the bears to hunt seals on. This reduces their access to food sources and threatens their habitat and overall survival.

6) Adélie penguins

These birds live in the Antarctic and feed on krill (found under the ice sheets).

As the ice melts, krill populations decrease and the penguins have to migrate from their natural habitat in an attempt to find alternative food sources. This influences their breeding patterns negatively.

June 2019

www.gvi.co.uk

Sea-level rise

Why is sea-level rise important?

Sea-level rise increases the frequency and severity of storm surges and coastal flooding, causing serious damage to critical infrastructure and leading to the displacement of coastal communities around the world. Globally, more than 100 million people live in coastal regions vulnerable to sea-level rise, and many of the world's largest cities are situated less than 10 metres above current sea level. In the UK, current annual damages from coastal flooding are estimated at over £500 million per year, and costs of damage are likely to increase under projections of future sea-level rise. As such, sea-level rise presents one of the biggest adaptation challenges to climate change.

How much and how fast is sea level rising?

Since 1900, global mean sea level (GMSL) has risen by approximately 20cm. The rate of sea-level rise has increased throughout the 20th and early 21st centuries, and it is currently rising at about 3.2cm per decade.

What causes sea level to rise?

Increasing global surface temperatures, resulting from human emissions of greenhouse gases (GHGs), cause the sea level to rise through two main processes. Firstly, more than 90% of the excess heat in the atmosphere is absorbed by the oceans, causing the ocean to increase in volume as it warms. Secondly, water that is currently stored on land in the form of ice is added to the oceans as glaciers and ice sheets melt, further increasing ocean volume.

How do the ice sheets contribute to sea-level rise?

Together, the ice sheets in Greenland and Antarctica hold over 99% of all ice on Earth. Observations show that the sea-level contribution due to ice loss from these ice sheets has tripled in the past two decades, and now accounts for a third of total GMSL rise. Enhanced ice loss from the ice sheets can occur either through increased melting of ice at their surfaces, with the resulting meltwater running off into the ocean, or from an acceleration of the rate of ice flow into the ocean. In Greenland, surface melting during summer months is the dominant driver of this ice loss, while in Antarctica the majority of ice loss is caused by a speed-up of ice-sheet flow in certain regions.

The flow of the Antarctic Ice Sheet towards the ocean is to a large extent controlled by the floating ice shelves that fringe its coastline, which act as buttresses to stem the flow.

Observations show that these ice shelves are experiencing widespread thinning, driven by melting both from above due to warmer temperatures and from below by warm ocean waters. This is particularly the case for ice shelves along the coast of the West Antarctic Ice Sheet (WAIS), where the largest accelerations in ice-sheet flow are also observed, especially for the Pine Island and Thwaites Glaciers.

There are also concerns that parts of the WAIS that rest on bedrock below sea level could be unstable. Thinning in coastal regions may cause acceleration of the ice sheet and further thinning upstream, making sections of the WAIS vulnerable to collapse. In contrast, the East Antarctic Ice Sheet, which rests largely on bedrock above sea level, appears to be more stable, although some parts of East Antarctica may also be vulnerable to oceanic melting.

How has sea level changed in the past?

There is evidence that during extended warm periods in the past (125,000 and 400,000 years ago) a large fraction of Greenland was ice-free and sea levels rose slowly over centuries to be more than 6m higher than today. Local conditions influence the ice sheets, but during these periods the global average temperature was perhaps only slightly warmer than today.

How much will sea level rise in future?

The recent Intergovernmental Panel on Climate Change (IPCC) Special Report on Global Warming of 1.5°C projects that GMSL rise by 2100 is likely to lie within the ranges of 22–77cm or 35–93cm for warming of 1.5°C or 2°C above pre-industrial levels, respectively. The largest source of uncertainty in these projections is currently associated with quantifying the potential additional sea-level rise contribution due to instability of the WAIS, which remains the subject of ongoing research. The recently launched International Thwaites Glacier Collaboration, a research programme involving US and UK scientists including from

	RCP2.6		RCP4.5		RCP8.5	
	5th	95th	5th	95th	5th	95th
LONDON	0.29	0.70	0.37	0.83	0.53	1.15
CARDIFF	0.27	0.69	0.35	0.81	0.51	1.13
EDINBURGH	0.08	0.49	0.15	0.61	0.30	0.90
LONDON	0.11	0.52	0.18	0.64	0.33	0.94

0.3 0.4 0.5 0.6 0.7

Sea-level change (m)

Range of sea-level change (m) at UK capital cities in 2100 relative to 1981–2000 average for a low (RCP2.6), medium (RCP4.5) and high (RCP8.5) emissions scenario.

Source: Palmer M, et al., 2018. UKCP18 Marine Report. Met Office.

BAS, aims to improve understanding of ice-sheet stability in this region and reduce these uncertainties in future sea-level projections.

Beyond 2100, sea level will continue to rise for many centuries, even if GHG emissions are reduced to net-zero in line with the 2016 Paris Agreement targets to limit global warming to 1.5°C or 2°C. However, the magnitude and the rate of this committed long-term sea-level rise depends strongly on near-term emissions reductions in coming decades. The sooner net-zero or net-negative GHG emissions are achieved, the more the amount of long-term sea-level rise can be limited. If GHG emissions are left unchecked, the rate of sea-level rise will further accelerate. The IPCC 1.5°C report stated that instabilities of the Greenland and West Antarctic Ice Sheets could be triggered between 1.5°C and 2°C of warming, which would eventually result in several metres of sea-level rise over hundreds or thousands of years. Some glaciologists consider such instabilities could be triggered even below this level, but there is little doubt that the greater the warming, the greater the likelihood of such events occurring.

How does sea-level rise vary locally?

Sea-level rise is not uniform globally, as it is affected by multiple local and non-local factors including gravitational effects, ocean circulation patterns and vertical land movement.

Projections of GMSL rise, including BAS estimates of the future Antarctic sea-level contribution, therefore feed into locally-specific sea-level rise projections, which are crucial for informing local adaptation planning. The Met Office recently published its updated UK sea-level projections as part of the UK Climate Projections 2018, which show how the amount of sea-level rise by 2100 will vary for different UK coastal locations and different GHG emission scenarios.

21 May 2019

Will climate change cause humans to go extinct?

THE CONVERSATION

***An article from* The Conversation.**

By Anders Sandberg, James Martin Research Fellow, Future of Humanity Institute & Oxford Martin School, University of Oxford

'I see a lot of resources talking about near-term human extinction, or the fact that thanks to climate change my generation will see the end of humanity. How likely is an outcome like this? Is there any hope for our futures?' **Anonymous, aged 18. London, UK.**

The claim that humanity only has just over a decade left due to climate change is based on a misunderstanding. In 2018, a fairly difficult-to-read report by the Intergovernmental Panel on Climate Change (IPCC) warned that humanity needs to cut its carbon dioxide (CO_2) emissions in half by 2030, to avoid global warming of 1.5°C above the levels seen before the Industrial Revolution.

What this actually means is roughly, 'We have about 12 years before fixing climate change becomes really expensive and tough.'

Humanity can still live in a world with climate change – it's just going to be more work, and many lives and livelihoods are likely to be threatened. But it's complicated, because this century we are facing many problems at the same time, and we are more dependent on each other than ever.

Under pressure

To get our food, most of us humans depend on global transport, payment and logistics systems. These, in turn, require fuel, electricity, communications and a lot of other things to work properly.

All these systems are connected to each other, so if one starts crashing, the chaos may cause other systems to crash, and before we know it we'll have massive shortages and conflicts.

It's hard to calculate the exact risk of this happening, since it has never happened before. Until recently, the world was split into separate regions that were largely independent of each other.

But we do know that climate change puts the whole world under pressure – everywhere, at the same time – making the risk of these systems collapsing more serious.

For example, it's easier for businesses to handle cybersecurity and energy supply when they don't also have to cope with natural hazards. Likewise, it's difficult for governments to maintain infrastructure when politicians are busy dealing with the public's reactions to food prices, refugees and ecological crises.

Building resilience

Geoengineering to reduce the impact of climate change – for example, by reducing CO_2 levels or pumping reflective particles into the Earth's atmosphere to deflect the sun's rays – might work. But if disaster strikes and those operations stop, the effects of climate change can return quickly.

The reasonable thing to do is to work on making our systems more resilient – and there are plenty of opportunities to do this. In practice, this means more local energy production, better backup systems, work on reducing climate change, and being more willing to pay extra for safety.

Disasters and diseases

So what about the other threats humanity is facing? Though natural hazards such as earthquakes, tsunamis, volcanoes and hurricanes can be disastrous, they pose a comparatively small threat to the survival of the human race.

Hazards big enough to cause entire species to go extinct are relatively rare. The typical mammalian species survives for about a million years, so the risk is roughly one in a million per year. Asteroid impacts and supervolcanos do happen, but they are rare enough that we do not have to worry about them. Even so, planning for the day when we need to deflect an asteroid or make do without agriculture for a decade is a smart move.

Pandemics are worse. We know the 1918 flu killed tens of millions of people worldwide. New influenza viruses are popping up all the time, and we should expect to see a big pandemic at least once every 100 years.

Over the past century, we have become better at medicine (which lowers the risk from disease) but we also travel more (which increases the spread of diseases). Natural pandemics are unlikely to wipe out the human race, since there is almost always somebody who is immune. But a bad pandemic might still wreck our global society.

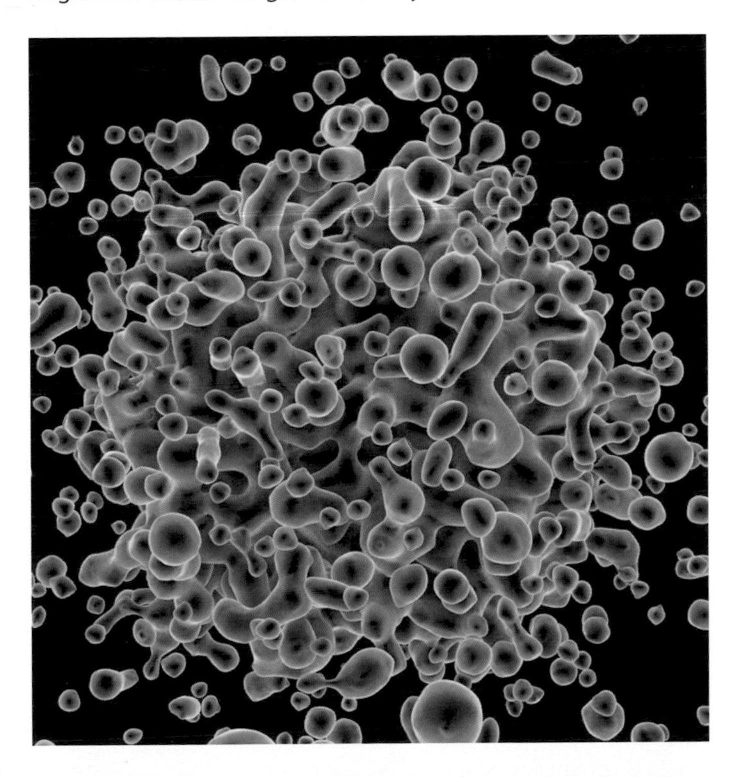

Technology attacks

Bioweapons, which use bacteria, viruses or fungi to harm humans or agriculture, are another issue. Fortunately, they have rarely been used in war, but they might become more dangerous in the near future because advances in biotechnology are making it easier and cheaper to modify organisms and automate lab work.

As this technology becomes more accessible, there's a growing risk it could be used as a 'doomsday device' by nasty regimes, to deter other states from seeking to topple them.

Right now, the risk is smallish, but it will surely become larger if we do not figure out better ways to detect pathogens early on, keep an eye on risky biotechnology and do diligent diplomacy to keep governments sane.

Perhaps the biggest risk to humanity right now is nuclear weapons. I would personally guess the risk of a nuclear war (not necessarily world-ending but still horrifying) to be somewhere between one in 100 and one in 1,000 per year. This risk goes up or down, depending on tensions between countries and the competence of the people handling early warning systems.

At the Future of Humanity Institute at the University of Oxford, we do a lot of work on Artificial Intelligence (AI). As with biotechnology, the risk right now is pretty minimal, but it might grow in time as AI become better and smarter, and we think it's better to be safe than sorry.

Developing tools to ensure AI stays safe and operates in a way that benefits humanity could save money in the long run, and it's unlikely to make things worse. Again, the probability of an AI disaster is fairly undefined, since it changes depending on how well we prepare for it.

I can't give a probability of a world-ending disaster that isn't more or less guesswork. But I do think there's a big enough risk of such a disaster in our lifetimes that we should work hard to fix the world – whether by making sure governments and AI stay safe and sane, replacing fossil fuels, building backup systems and plans, decentralising key systems and so on. These things are worthwhile, even if the risk is one in a million: the world is precious, and the future we are risking is vast.

29 May 2019

This article is part of *'I Need To Know,'* a Q&A service for teenagers by The Conversation.
If you're a teenager aged 12 to 18, and you've got questions you'd like an expert to answer, send them our way! Include your first name, age and the area you live in. To get in touch, you can:

- email ineedtoknow@theconversation.com
- ask a question using Incogneato
- DM us on Instagram @theconversationdotcom

We have a huge pool of experts at our fingertips, and we can't wait to share their knowledge with you.

Six pressing questions about beef and climate change, answered

By *Richard Waite, Associate, WRI's Food Program; Tim Searchinger, Senior Fellow, World Resources Report,* Creating A Sustainable Food Future, *World Resources Institute and Janet Ranganathan, Vice President For Science And Research, World Resources Institute*

Beef and climate change are in the news these days, from cows' alleged high-methane farts (fact check: they're actually mostly high-methane burps) to comparisons with cars and airplanes (fact check: the world needs to reduce emissions from fossil fuels and agriculture to sufficiently rein in global warming). And as with so many things in the public sphere lately, it's easy for the conversation to get polarized. Animal-based foods are nutritious and especially important to livelihoods and diets in developing countries, but they are also inefficient resource users. Beef production is becoming more efficient, but forests are still being cut down for new pasture. People say they want to eat more plants, but meat consumption is still rising.

All of the above statements are true even if they seem contradictory. That's what makes the beef and sustainability discussion so complicated – and so contentious.

Here we look at the latest research (including from our recent World Resources Report) to address six common questions about beef and climate change:

1. How does beef production cause greenhouse gas emissions?

The short answer: Through the agricultural production process and through land-use change.

The longer explanation: Cows and other ruminant animals (like goats and sheep) emit methane, a potent greenhouse gas, as they digest grasses and plants. This process is called 'enteric fermentation,' and it's the origin of cows' burps. Methane is also emitted from manure, and nitrous oxide, another powerful greenhouse gas, is emitted from ruminant wastes on pastures and chemical fertilizers used on crops produced for cattle feed.

More indirectly but also importantly, rising beef production requires increasing quantities of land. New pastureland is often created by cutting down trees, which releases carbon dioxide stored in forests.

A 2013 study by the UN Food and Agriculture Organization (FAO) estimated that total annual emissions from animal agriculture (production emissions plus land-use change) were about 14.5% of all human emissions, of which beef contributed 41%. That means emissions from beef production are roughly on par with those of India. Because FAO only modestly accounted for land-use-change emissions, this is a conservative estimate.

Beef-related emissions are also projected to grow. Building from an FAO projection, we estimated that global demand for beef and other ruminant meats could grow by 88% between 2010 and 2050, putting enormous pressure on forests, biodiversity and the climate. Even after accounting for continued improvements in beef production efficiency, pastureland could still expand by roughly 400 million hectares, an area of land larger than the size of India, to meet growing demand. The resulting deforestation could increase global emissions enough to put the global goal of limiting temperature rise to 1.5–2 degrees C (2.7–3.6 degrees F) out of reach.

2. Is beef more resource-intensive than other foods?

The short answer: Yes.

The longer explanation: Ruminant animals have lower growth and reproduction rates than pigs and poultry, so they require a higher amount of feed per unit of meat produced. Animal feed requires land to grow, which has a carbon cost associated with it, as we discuss below.

All told, beef is more resource-intensive to produce than most other kinds of meat, and animal-based foods overall are more resource-intensive than plant-based foods. Beef requires 20 times more land and emits 20 times more GHG emissions per gram of edible protein than common plant proteins, such as beans. And while the majority of the world's grasslands cannot grow crops or trees, such 'native grasslands' are already heavily used for livestock production, meaning additional beef demand will likely increase pressure on forests.

Beef is more resource intensive than most other food

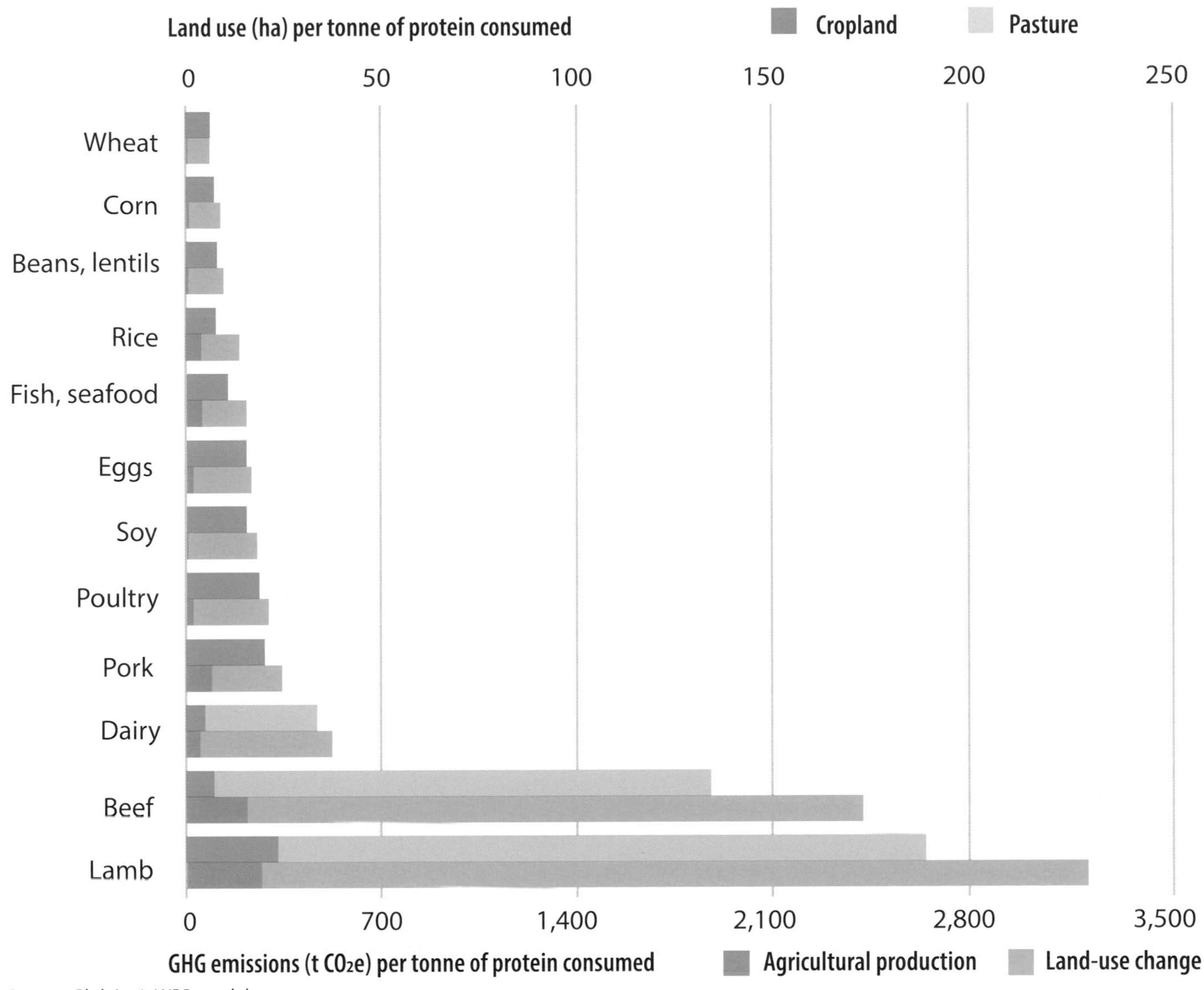

Source: GlobAgri -WRR model

3. Why are some people saying beef production is only a small contributor to emissions?

The short answer: Such estimates commonly leave out land-use impacts, such as cutting down forests to establish new pastureland.

The longer explanation: There are a lot of statistics out there that account for emissions from beef production but not from associated land-use change. For example, here are three common US estimates we hear:

- The US Environmental Protection Agency estimated total US agricultural emissions in 2017 at only 8% of total US emissions;
- A 2019 study in Agricultural Systems estimated emissions from beef production at only 3% of total US emissions; and
- A 2017 study published in the *Proceedings of the National Academy of Sciences* estimated that removing all animals from US agriculture would reduce US emissions by only 3%.

While all of these estimates account for emissions from US agricultural production, they leave out a crucial element: emissions associated with devoting land to agriculture. An acre of land devoted to food production is often an acre that could store far more carbon if allowed to grow forest or its native vegetation. And when considering the emissions associated with domestic beef production, you can't just look within national borders, especially since global beef demand is on the rise. Because food is a global commodity, what is consumed in one country can drive land use impacts and emissions in another. An increase in US beef consumption, for example, can result in deforestation to make way for pastureland in Latin America. Conversely, a decrease in US beef consumption can avoid deforestation (and land-use-change emissions) abroad.

When these land-use effects of beef production are accounted for, we found that the GHG impacts associated with the average American-style diet actually come close to per capita US energy-related emissions. A related analysis found that the average European's diet related emissions, when accounting for land-use impacts, are similar to the per capita emissions typically assigned to each European's consumption of all goods and services, including energy.

4. Can beef be produced more sustainably?

The short answer: Yes, although beef will always be resource-intensive to produce.

The longer explanation: The emissions intensity of beef production varies widely across the world, and improvements in the efficiency of livestock production can greatly reduce land use and emissions per pound of meat. Improving feed quality and veterinary care, raising improved animal breeds that convert feed into meat and milk more efficiently, and using improved management practices like rotational grazing can boost productivity and soil health while reducing emissions. Boosting productivity, in turn, can take pressure off tropical forests by reducing the need for more pastureland.

Examples of such improved practices abound. For example, some beef production in Colombia integrates trees and grasses onto pasturelands, helping the land produce a higher quantity and quality of feed. This can enable farmers to quadruple the number of cows per acre while greatly reducing methane emissions per pound of meat, as the cows grow more quickly. A study of dairy farms in Kenya found that supplementing typical cattle diets with high-quality feeds like napier grass and high-protein Calliandra shrubs – which can lead to faster cattle growth and greater milk production – could reduce methane emissions per litre of milk by 8–60%.

There are also emerging technologies that can further reduce cows' burping, such as through feed additives like 3-nitrooxypropan (3-NOP). Improving manure management and using technologies that prevent nitrogen in animal waste from turning into nitrous oxide can also reduce agricultural emissions.

5. Do we all need to stop eating beef in order to curb climate change?

The short answer: No.

The longer explanation: Reining in climate change won't require everyone to become vegetarian or vegan, or even to stop eating beef. If ruminant meat consumption in high consuming countries declined to about 50 calories a day or 1.5 burgers per person per week – about half of current US levels and 25% below current European levels, but still well above the national average for most countries – it would nearly eliminate the need for additional agricultural expansion (and associated deforestation), even in a world with 10 billion people.

Limiting ruminant meat consumption to less than 50 calories/person/day by 2050 nearly eliminates the need for agricultural land expansion

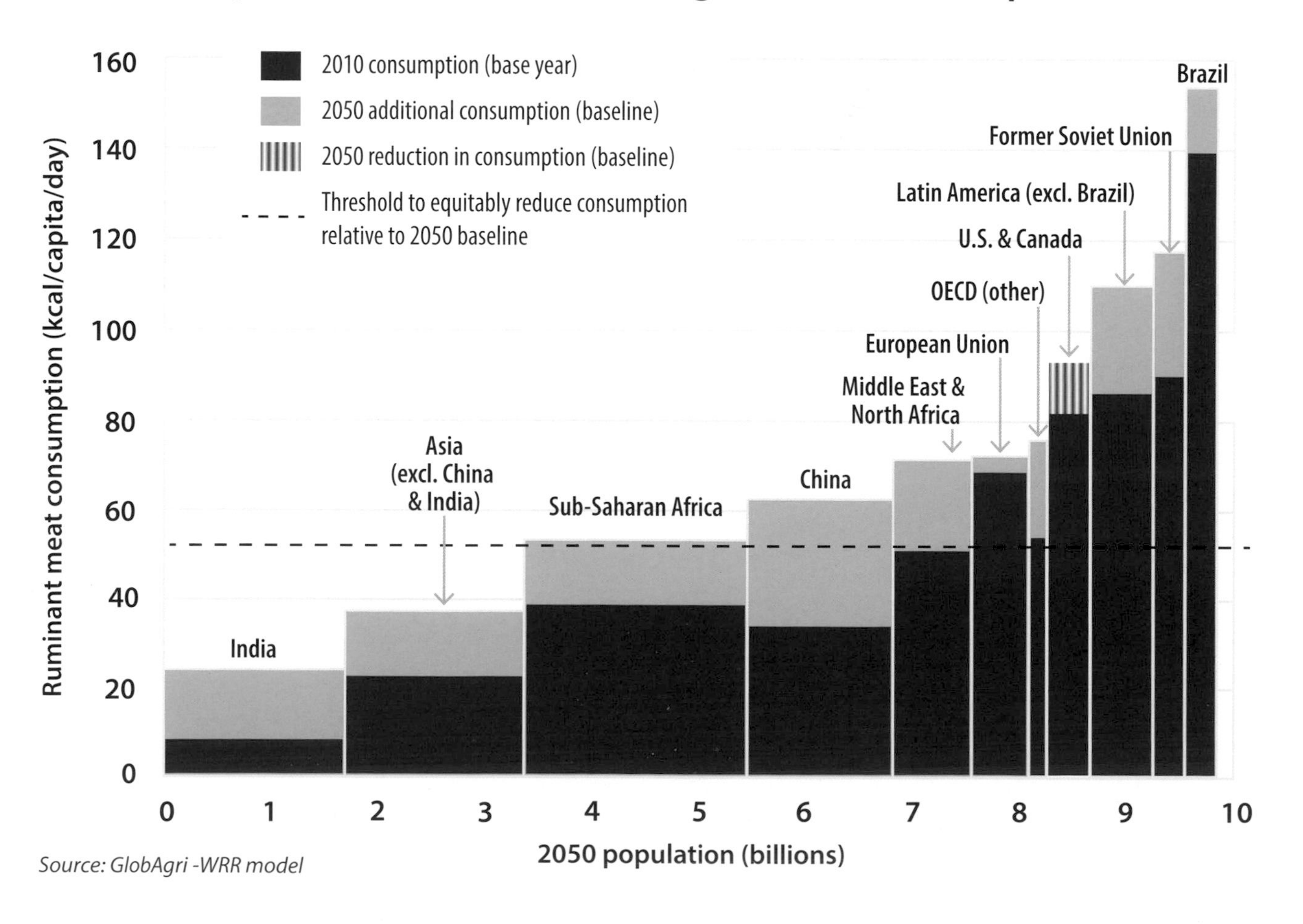

Source: GlobAgri -WRR model

Diets are already shifting away from beef in some places. Per capita beef consumption has already fallen by one-third in the United States since the 1970s. Plant-based burgers and blended meat-plant alternatives are increasingly competing with conventional meat products on important attributes like taste, price and convenience. The market for plant-based alternatives is growing at a high rate, albeit from a low baseline.

There are also other compelling reasons for people to shift towards plant-based foods. Some studies have shown that red meat consumption is associated with increased risk of heart disease, type 2 diabetes, stroke and colorectal cancer, and that diets higher in healthy plant-based foods (such as whole grains, fruits, vegetables, nuts and legumes) are associated with lower risks. In high-income regions like North America and Europe, people also consume more protein than they need to meet their dietary requirements.

6. Would eating less beef be bad for jobs in the food and agriculture sector?

The short answer: Not necessarily.

The longer explanation: Given projected future growth in meat demand across the developing world, even if people in higher-income countries eat less beef, the global market for beef will likely continue to grow in the coming decades. The scenario in the chart on page 26 leads to a 32% growth in global ruminant meat consumption between 2010 and 2050, versus 88% growth under business-as-usual. In the US, despite declining per capita beef consumption, total beef production has held steady since the 1970s. Burgeoning demand in emerging markets like China will lead to more export opportunities in leading beef-producing countries, although building such markets takes time.

In addition, major meat companies – including Tyson Foods, Cargill, Maple Leaf Foods and Perdue – are starting to invest in the fast-growing alternative protein market. They're positioning themselves more broadly as "protein companies," even as they work to reduce emissions from beef production in their supply chains through improved production practices.

Moving toward a sustainable food future

Beef is more resource-intensive than most other foods and has a substantial impact on the climate. A sustainable food future will require a range of strategies from farm to plate. Food producers and consumers alike have a role to play in reducing beef's emissions as the global population continues to grow. And as we all work on strategies to curb climate change – whether in the agriculture sector, the energy sector or beyond – it's important we rely on the best available information to make decisions.

15 April 2019

www.wri.org

Halfway to boiling: the city at 50°C

By Jonathan Watts

Imagine a city at 50°C (122°F). The pavements are empty, the parks quiet, entire neighbourhoods appear uninhabited. Nobody with a choice ventures outside during daylight hours. Only at night do the denizens emerge, H.G. Wells-style, into the streets – though, in temperatures that high, even darkness no longer provides relief. Uncooled air is treated like effluent: to be flushed as quickly as possible.

School playgrounds are silent as pupils shelter inside. In the hottest hours of the day, working outdoors is banned. The only people in sight are those who do not have access to air conditioning, who have no escape from the blanket of heat: the poor, the homeless, undocumented labourers. Society is divided into the cool haves and the hot have-nots.

Those without the option of sheltering indoors can rely only on shade, or perhaps a water-soaked sheet hung in front of a fan. Construction workers, motor-rickshaw drivers and street hawkers cover up head to toe to stay cool. The wealthy, meanwhile, go from one climate-conditioned environment to another: homes, cars, offices, gymnasiums, malls.

Asphalt heats up 10–20°C higher than the air. You really could fry an egg on the pavement. A dog's paws would blister on a short walk, so pets are kept behind closed doors. There are fewer animals overall; many species of mammals and birds have migrated to cooler environments, perhaps at a higher altitude – or perished. Reptiles, unable to regulate their body temperatures or dramatically expand their range, are worst placed to adapt. Even insects suffer.

Maybe in the beginning, when it was just a hot spell, there was a boom in spending as delighted consumers snapped up sunglasses, bathing suits, BBQs, garden furniture and beer. But the novelty quickly faded when relentless sunshine became the norm. Consumers became more selective. Power grids are overloaded by cooling units. The heat is now a problem.

The temperature is recalibrating behaviour. Appetites tend to fade as the body avoids the thermal effect of food and tempers are quicker to flare – along, perhaps, with crime and social unrest. But eventually lethargy sets in as the body shuts down and any prolonged period spent outdoors becomes dangerous.

Hospitals see a surge in admissions for heat stress, respiratory problems and other illnesses exacerbated by high temperatures. Some set up specialist wards. The elderly, the obese and the sick are most at risk. Deaths rise.

At 50°C – halfway to water's boiling point and more than 10°C above a healthy body temperature – heat becomes toxic. Human cells start to cook, blood thickens, muscles lock around the lungs and the brain is choked of oxygen. In dry conditions, sweat – the body's inbuilt cooling system – can lessen the impact. But this protection weakens if there is already moisture in the air.

A so-called 'wet-bulb temperature' (which factors in humidity) of just 35°C can be fatal after a few hours to even the fittest person, and scientists warn climate change will make such conditions increasingly common in India, Pakistan, south-east Asia and parts of China. Even under the most optimistic predictions for emissions reductions, experts say almost half the world's population will be exposed to potentially deadly heat for 20 days a year by 2100.

Not long ago, 50°C was considered an anomaly, but it is increasingly widespread. Earlier this year, the 1.1 million residents of Nawabshah, Pakistan, endured the hottest April ever recorded on Earth, as temperatures hit 50.2°C. In neighbouring India two years earlier, the town of Phalodi sweltered in 51°C – the country's hottest ever day.

Dev Niyogi, professor at Purdue University, Indiana, and chair of the Urban Environment department at the American Meteorological Society, witnessed how cities were affected by extreme heat on a research trip to New Delhi and Pune during that 2015 heatwave in India, which killed more than 2,000 people.

'You could see the physical change. Road surfaces started to melt, neighbourhoods went quiet because people didn't go out and water vapour rose off the ground like a desert mirage,' he recalls.

'We must hope that we don't see 50°C. That would be uncharted territory. Infrastructure would be crippled and ecosystem services would start to break down, with long-term consequences.'

Several cities in the Gulf are getting increasingly accustomed to such heat. Basra – population 2.1 million – registered 53.9°C two years ago. Kuwait City and Doha have experienced 50°C or more in the past decade. At Quriyat, on the coast of Oman, overnight temperatures earlier this summer remained above 42.6°C, which is believed to be the highest 'low' temperature ever recorded in the world.

At Mecca, the two million hajj pilgrims who visit each year need ever more sophisticated support to beat the heat. On current trends, it is only a matter of time before temperatures exceed the record 51.3°C reached in 2012. Last year, traditionalists were irked by plans to install what are reportedly the world's biggest retractable umbrellas to provide shade on the courtyards and roof of the Great Mosque. Air conditioners weighing 25 tonnes have been brought in to ventilate four of the biggest tents. Thousands of fans already cool the marble floors and carpets, while police on horseback spray the crowds with water.

Football supporters probably cannot expect such treatment at the Qatar World Cup in 2022, and many may add to the risks of hyperthermia and dehydration by taking off their shirts and drinking alcohol. FIFA is so concerned about conditions that it has moved the final from summer to a week before Christmas. Heat is also why Japanese politicians are now debating whether to introduce daylight saving time for the 2020 Tokyo Olympics so that marathon and racewalk athletes can start at what is currently 5am and avoid mid-afternoon temperatures that recently started to pass 40°C with humidity of more than 80%.

At the Australian open in Melbourne this year – when ambient temperatures reached 40°C – players were staggering around like 'punch-drunk boxers' due to heatstroke. Even walking outside can feel oppressive at higher temperatures. 'The blast of furnace-like heat ... literally feels life-threatening and apocalyptic,' says Nigel Tapper, professor of environmental science at Melbourne's Monash University, of the 48°C recorded in parts of the city. 'You cannot move outside for more than a few minutes.'

The feeling of foreboding is amplified by the increased threat of bush and forest fires, he adds. 'You cannot help but ask, "How can this city operate under these conditions? What can we do to ensure that the city continues to provide important services for these conditions? What can we do to reduce temperatures in the city?"'

Those places already struggling with extreme heat are doing what they can. In Ahmedabad, in Gujarat, hospitals have opened specialist heat wards. Australian cities have made swimming pools accessible to the homeless when the heat creeps above 40°C, and instructed schools to cancel playground time. In Kuwait, outside work is forbidden between noon and 4pm when temperatures soar.

But many regulations are ignored, and companies and individuals underestimate the risks. In almost all countries, hospital admissions and death rates tend to rise when temperatures pass 35°C – which is happening more often, in more places. Currently, 354 major cities experience average summer temperatures in excess of 35°C; by 2050, climate change will push this to 970, according to the recent *Future We Don't Want* study by the C40 alliance of the world's biggest metropolises. In the same period, it predicts the number of urban dwellers exposed to this level of extreme heat will increase eightfold, to 1.6 billion.

As baselines shift across the globe, 50°C is also uncomfortably near for tens of millions more people. This year, Chino, 50km (30 miles) from Los Angeles, hit a record of 48.9°C, Sydney saw 47°C, and Madrid and Lisbon also experienced temperatures in the mid-40s. New studies suggest France 'could easily exceed' 50°C by the end of the century while Australian cities are forecast to reach this point even earlier. Kuwait, meanwhile, could sizzle towards an uninhabitable 60°C.

How to cool dense populations is now high on the political and academic agenda, says Niyogi, who last week co-chaired an urban climate symposium in New York. Cities can be modified to deplete heat through measures to conserve water, create shade and deflect heat. In many places around the world, these steps are already under way. The city at 50°C could be more tolerable with lush green spaces on and around buildings; towers with smart shades that follow the movement of the sun; roofs and pavements painted with high-albedo surfaces; fog capture and renewable energy fields to provide cooling power without adding to the greenhouse effect.

But with extremes creeping up faster than baselines, Niyogi says this adapting will require changes not just to the design of cities, but how they are organised and how we live in them. First, though, we have to see what is coming – which might not hit with the fury of a flood or typhoon but can be even more destructive.

'Heat is different,' says Niyogi. 'You don't see the temperature creep up to 50°C. It can take people unawares.'

13 August 2018

A Leicester fish and chip shop is closing because its owners are worried about the environment

Aatin and Helen Anadkat served fish and chips cooked in rapeseed oil with a gluten-free batter as well as dairy-free and vegan dishes.

By Dean Kirby

The owners of a fish and chip shop have closed it down after saying they had begun to feel uncomfortable about the environmental impact of their business.

Aatin and Helen Anadkat opened The Fish and The Chip restaurant in Leicester in 2017 in a bid to bring a modern twist to traditional fish suppers.

They served fish and chips cooked in rapeseed oil with a gluten-free batter as well as a range of dairy-free and vegan dishes.

But now the couple have decided to shut down the restaurant after spending months considering the business's impact.

Declining fish numbers

In a sign posted in the restaurant's window, they said: 'We saw the impact pollution is having on the oceans and fish stocks and we are not comfortable running a restaurant that has an impact on our environment.

'As a result, we have decided to close the restaurant.'

Mr Anadkat told ***i*** his main concerns were around declining fish numbers and fish eating plastic particles in the water.

He said: 'We launched the restaurant in 2017 and it grew and grew. We did really well. But the fish didn't sit with us really well.

'I don't want to be seen to be having a go at the fishing industry, but fish stocks are declining and moving further afield to fish destroys new eco-systems.

'I'm also concerned about plastic pollution in the oceans and how that is affecting the fish and affecting what we are putting on our plates.

'Over time, we came to the conclusion that we wanted to close the business.'

Overfishing has been described as one of the most significant drivers of the decline in ocean wildlife by the WWF, with scientists saying last month that North Sea cod stocks have fallen to critical levels.

The International Council for the Exploration of the Sea has said cod – which along with haddock is the staple of fish suppers – is being harvested 'unsustainably'.

'Their future is in our hands'

Mr and Mrs Anadkat are now opening a sustainable and environmentally-friendly vegan food manufacturing business called Positive Kitchen, which will be dairy-free, gluten-free and contain no refined sugar or genetically-modified ingredients.

Mr Anadkat said: 'It's a risk but it's one we're prepared to take. With this company, we can influence so many more people around the country and start to be a national movement.

'My youngest son has a dairy allergy and it make you think. Their future is in our hands and we need to make changes to protect the environment now rather than in the future.

'It just seems that the current system of food is a bit broken and would be nice to help fix it.'

9 July 2019

Chapter 3 **Action**

Extinction Rebellion did all it could have hoped and more – but our crushing inertia remains its biggest obstacle

It is fair to conclude that the wave of direct action in the capital, turbo-charged by the arrival of Greta Thunberg, will bring about no immediate change in policy.

By Sean O'Grady

Did Extinction Rebellion do much more than annoy London taxi drivers and right-wing climate change denying columnists?

It is fair to conclude that the wave of direct action in the capital, its moral force turbo-charged by the arrival of the 16-year-old campaigner Greta Thunberg, will bring about no immediate change in policy. There will be no emergency legislation to further restrict carbon dioxide emissions. A timely series of warnings in TV programmes presented by that other secular saint, Sir David Attenborough has added another unimpeachable voice to the cause.

Even so, the price of petrol will remain where it is. The long-haul flights will still depart from Heathrow. The environment secretary, Michael Gove has said that the government 'hears' the cry for change. That is, for the time being, as good as it gets.

More to the point, China will still be building new power stations, and the US will remain outside the structures of the Paris Accord on climate change. Deforestation, desertification, species extinctions and extreme weather events will continue to intensify.

And yet these protests have not been in vain. All the progress that has been made over the last 50 years on international action over climate change has been a product of popular movements, themselves spurred by the weight of scientific evidence and the leadership of – some – politicians.

More to the point, China will still be building new power stations, and the US will remain outside the structures of the Paris Accord on climate change. Deforestation, desertification, species extinctions and extreme weather events will continue to intensify.

From the first awakenings in the 1970s, the green movement has become ever more powerful. Like the suffragettes and the civil rights campaigns before them, and like those for the dismantling of colonial empires and LGBT+ rights, movements once ridiculed can bring change that is radical and permanent. Like those other successful incremental past movements the environmental cause has made itself increasingly mainstream.

Today, even after the efforts of Donald Trump and a global gang of client deniers, much of the world has mandatory targets on climate change and CO_2 emissions – with Britain leading the way. The internal combustion engine will become a museum piece like the steam engine by mid-century. Where once virtually every machine, every train, every ship, and every power station in Britain was fuelled by coal, the UK has just enjoyed its longest period of zero coal derived electricity generation ever – 90 hours.

Now China is leading the world in electric propulsion and battery power. More and more cities want to charge for pollution and congestion. Citizens are clamouring for change, especially in the West.

Yet there is a great disconnect, a cognitive dissonance For once, the TV vox pops from the Extinction Rebellion were enlightening on this point: even many of those inconvenienced by the protests were sympathetic to the arguments. Few doubt what we are doing to the planet is suicidal. In a sense these protesters, unlike the average striker say, enjoy mass public support. But when the moment comes to book a week in Bali or buy a new car or go shopping for food swaddled in plastics we seem to forget our personal responsibilities. When we go to vote we rarely put the planet where it logically ought to be – at the top of our agenda.

Unlike Brexit, for example, or terror, the dangers of climate change seem entirely remote in time and geography – rainforests and Pacific islands completely disappearing over a period of decades rather than months, but nothing tangible nearer home. The science can sound simultaneously forbidding and trivial – a one degree rise in sea temperatures will have devastating consequences, but it is superficially inconsequential.

Gradually, perhaps, that is changing. There is more public understanding of the science. There is more acceptance of charges for congestion, for taxes on waste and pollution and support for investing in renewable energy. That is a very different climate of opinion from a few years ago.

Over Easter we have witnessed some bizarre dancing in the street, a pink boat marooned on Oxford Street, people glued to trains and Jeremy Corbyn's garden fence, and outbreaks of the curious 'jazz hands' phenomenon. We are talking about all that – and about climate change. Eccentric as much of it was, the protesters have drawn a little more attention to the immediacy of the threats the world faces. Getting out there was worth it.

26 April 2019

www.independent.co.uk

Climate change: 10 easy ways to help plants and animals from becoming extinct

Climate change, urban development, air pollution and intensive farming have all taken their toll on wildlife.

By Tom Bawden

Climate change, urban development, air pollution, and intensive farming have all taken their toll on wildlife. Whether it's plastic bags or pesticides, every technological advance seems to come with a price paid by animals and plants.

One recent report, by the United Nations, warned that a million species are at risk of extinction because of the actions of people.

Such reports – and there are plenty – can create a sense of helplessness.

Nevertheless, there are ways that each of us can make a difference without having to transform into an all-or-nothing eco-warrior.

The United Nations has called on citizens to do what they can to preserve the number and diversity of wildlife species. Here are some of the simple ways we can – and many of them already do – help.

Put in a bird feeder

Up to half of UK householders are now thought to feed birds in their gardens, providing an estimated 60,000 thousand tonnes of bird food a year.

The ideal bird feeder is sturdy enough to withstand winter weather and squirrels, tight enough to keep seeds dry, easy to assemble and, most important of all, easy to keep clean.

They come in all sorts of shapes and sizes, from trays and tubes, to suit locations from window ledges to trees.

And it's not just food – from kitchen scraps to nuts threaded on string – that is a draw.

While the days when feeding garden birds was centred on kitchen scraps, suet-filled coconut shells and monkey nuts threaded on string are largely gone, those remain acceptable, but these days many people avail of the multi-million pound wild-bird food industry, which offers a bewildering array of food and feeders. These days, says Dr Kate Plummer, of the British Trust for Ornithology, you can buy a feeder that 'also provides water and places for animals to shelter and breed'.

Let your lawn grow free

While many gardeners prize a well-maintained lawn, conservationists say it's far better for nature to leave your mower in the shed.

Wildflower-studded lawns are an increasingly important source of nectar for pollinators such as bees and butterflies since nearly 7.5 million acres of meadows and pastures rich in wildflowers have been lost since the 1930s, removing a vital source of food for insects, many of which are in decline.

This is a major issue, as one acre of wildflower meadow on a single day in summer can contain three million flowers, producing 1kg of nectar – enough to support nearly 96,000 honeybees per day, according to the insect charity Plantlife.

As meadows decline, it falls to Britain's 15 million gardens many of them with lawns, to take up the slack and become an increasingly important habitat to help pollinators in search of food through flowers such as common daisies, red clover, dandelions, dove's foot, cranes bill and buttercups.

The insect charity Plantlife launched a campaign last month (May) to encourage homeowners to go easy on their lawns.

'Our research shows that people really want to do their bit to help pollinators and one of the best ways to do that is to give the wildflowers in our lawns a chance to flower. It might sound wrong but by doing nothing to your lawn you will actually help nature,' said Plantlife's Trevor Hines.

Volunteer to record wildife

The Big Butterfly Count will take place from 19 July to 11 August. Butterflies react very quickly to changes in their environment, which makes them excellent biodiversity indicators. Butterfly declines are an early warning of other wildlife losses.

That's why counting butterflies can be described as taking the pulse of nature.

The count will also help to identify trends that will help us plan how to protect butterflies from extinction, as well as understand the effect of climate change on wildlife.

To take part, you simply count butterflies for 15 minutes during bright (preferably sunny) weather during the period. Then send in your sightings online at bigbutterflycount.org or by using free big butterfly count dedicated smartphone apps available for iOS and Android.

You could also help count hedgehogs in Regent's Park for two nights in May and two nights in September – contact volunteering@royalparks.org.uk to register.

And the RSPB does a Big Garden Birdwatch at the end of January each year, while the Woodland Trust is looking for 'official wildlife recorders'.

Help in the battle against climate change

Climate change is already having a huge effect on wildlife around the world and we have had only a relatively small 1°C of warming since pre-industrial times. The fear is that without drastic action, warming could rise to 3°C or 4°C by the end of the century and maybe even higher. That being the case, people could make a huge contribution of the plight of nature by cutting energy use as much as possible – buying energy-efficient appliances, turning off lights when not needed, running full washing machine loads at cooler temperatures, unplugging electrical appliances when not in use to reduce standby power use and insulating their houses.

And you can help in another way – by switching to an energy provider that generates all of its electricity from low-carbon nuclear or renewable sources.

Grow your own fruit and veg

Leading bee expert and horticulturalist Professor Dave Goulson calculated for ***i*** last month that if UK households went on a growing spree, they could produce enough fruit and veg in their gardens to feed the whole country. Giving over half of the average garden to crops would produce 7.5 million tons of fruit and veg a year in the UK – more than the national consumption of 6.9 million tons.

Although that is a rather lofty aim, there is clearly scope for households to grow far more of the produce they eat. This is good for biodiversity, as the crops attract a range of pollinators and other species. But it is also good for our health – because it is likely to increase our consumption of organic fruit and vegetables and because the experience of growing them is therapeutic. It also saves money and would increase the country's food security.

Make a hole in the fence for hedgehogs

Putting a hole in the fence might not be for everyone, but experts say it can contribute to a 'hedgehog highway'.

Hedgehogs can travel up to a mile in a night, which means that fences and walls are getting in the way of their plans. By connecting as many gardens and green spaces as possible, the highway gives beleaguered hedgehogs more places to feed, nest or meet their partners. According to the Hedgehog Street campaign, making a hole is really easy to do (with your neighbour's permission of course). You just need to drill a 13cm x 13cm hole in your fence (about the size of a CD case), and this helps to prevent larger animals such as pets getting through, but is just the right size for hedgehogs.

Help in the battle against plastic pollution

A truckload of plastic waste is finding its way into the oceans every minute, where it breaks down and infiltrates every level of the food chain – eventually finding its way back on to the dinner plate. This plastic is wreaking havoc on marine wildlife and there are plenty of ways we can help. Coffee cups and plastic bottles are major sources of pollution, so you could switch to refillable cups. Throwaway plastic bags are another problem; so you could switch to cloth or other

reusable bags. Avoid fruit and veg that is unnecessarily wrapped and steer clear of wet wipes with high plastic content.

Create a pond

The number of ponds has declined significantly in recent decades, as land has been drained for farming or development. Yet ponds typically accommodate a much broader range of species than land, and even a small pond, as little as one square metre, can do wonders for your garden's biodiversity.

Among other things, ponds support dragonflies, aquatic beetles, mayflies, caddis flies and amphibians such as the great crested newt.

If you don't have room for a pond, a damp ditch can be an easy way to bring a little moisture to your garden, says the RSPB. Although it may not retain water like a pond, It will naturally collect rain and form a lovely home for all sorts of wildlife.

Don't pave over grass

If you're thinking of concreting over your garden or making a driveway, think again. Concrete or tarmac can accommodate very little in the way of wildlife compared with earth and ponds.

Don't use pesticides or fertilisers in your garden

Enormous 'monoculture' farms rely on pesticides because the scale of the crop attracts pests in droves, while there are no typically no natural predators to take them out because there is nowhere for them to live. But that is not the case for gardens, which are much more varied in terms of plants. This means that pests are not attracted in anything like the numbers, while there are usually predators that can control their numbers.

Pesticides cause considerable harm to wildlife and should be avoided where possible – as should synthetic fertilisers, which disrupt the soil's make-up. Making and using compost is an easy alternative.

14 June 2019

www.inews.co.uk

Net-zero – the UK's contribution to stopping global warming

An extract from a report by the Committee on Climate Change.

This report responds to a request from the Governments of the UK, Wales and Scotland, asking the Committee to reassess the UK's long-term emissions targets. Our new emissions scenarios draw on ten new research projects, three expert advisory groups, and reviews of the work of the IPCC and others.

The conclusions are supported by detailed analysis published in the *Net Zero Technical Report* that has been carried out for each sector of the economy, plus consideration of F-gas emissions and greenhouse gas removals.

The report's key findings are that:

- The Committee on Climate Change recommends a new emissions target for the UK: net-zero greenhouse gases by 2050.
- In Scotland, we recommend a net-zero date of 2045, reflecting Scotland's greater relative capacity to remove emissions than the UK as a whole.
- In Wales, we recommend a 95% reduction in greenhouse gases by 2050.

A net-zero GHG target for 2050 will deliver on the commitment that the UK made by signing the Paris Agreement. It is achievable with known technologies, alongside improvements in people's lives, and within the expected economic cost that Parliament accepted when it legislated the existing 2050 target for an 80% reduction from 1990.

However, this is only possible if clear, stable and well-designed policies to reduce emissions further are introduced across the economy without delay. Current policy is insufficient for even the existing targets.

How can the UK reach net-zero GHGs?

Scenarios for UK net-zero GHGs in 2050

It is impossible to predict the exact mix of technologies and behaviours that will best meet the challenge of reaching net-zero GHG emissions, but our analysis in this report gives an improved understanding of what a sensible mix might look like. Including:

- **Resource and energy efficiency** that reduce demand for energy across the economy. Without these measures the required amounts of low-carbon power, hydrogen and carbon capture and storage (CCS) would be much higher. In many, though not all, cases they reduce overall costs.
- Some **societal choices** that lead to a lower demand for carbon-intensive activities, for example an acceleration in the shift towards healthier diets with reduced consumption of beef, lamb and dairy products.
- Extensive **electrification**, particularly of transport and heating, supported by a major expansion of renewable and other low-carbon power generation. The scenarios involve around a doubling of electricity demand, with all power produced from low-carbon sources (compared to 50% today). That could for example require 75 GW of offshore wind in 2050, compared to 8 GW today and 30 GW targeted by the Government's sector deal by 2030. 75 GW of offshore wind would require up to 7,500 turbines and could fit within 1–2% of the UK seabed, comparable to the area of sites already leased for wind projects by the Crown Estate.
- Development of a **hydrogen** economy to service demands for some industrial processes, for energy-dense applications in long-distance HGVs and ships, and for electricity and heating in peak periods. By 2050, a new low-carbon industry is needed with UK hydrogen production capacity of comparable size to the UK's current fleet of gas-fired power stations.
- **Carbon capture and storage** (CCS) in industry, with bioenergy (for GHG removal from the atmosphere), and very likely for hydrogen and electricity production. CCS is a necessity not an option. The scenarios involve aggregate annual capture and storage of 75–175 $MtCO_2$ in 2050, which would require a major CO_2 transport and storage infrastructure servicing at least five clusters and with some CO_2 transported by ships or heavy goods vehicles.
- Changes in the way we farm and use our **land** to put much more emphasis on carbon sequestration and biomass production. Enabled by healthier diets and reductions in food waste, our scenarios involve a fifth of UK agricultural land shifting to tree planting, energy crops and peatland restoration.

Taken together, these measures would reduce UK emissions by 95–96% from 1990 to 2050. Tackling the remaining 4–5% would require some use of options that currently appear more speculative. That could involve greater shifts in diet and land use alongside more limited aviation demand growth, a large contribution from emerging technologies to remove CO_2 from the atmosphere (e.g. 'direct air capture'), or successful development of a major supply of carbon-neutral synthetic fuels (e.g. produced from algae or renewable power).

The scenarios involve additional reductions in the UK's consumption emissions as they include measures like resource efficiency that cut emissions from production overseas as well as in the UK. However, consumption emissions will only reach net-zero once the rest of the world's territorial emissions are also reduced to net-zero. At this point the UK can expect to pay slightly more to cover the costs of low-carbon production of the goods we import.

UK net-zero GHG scenario

	2020s	2030s and 2040s
Electricity	Largely decarbonise electricity: renewables, flexibility, coal phase-out	Expand electricity system, decarbonise mid-merit/ peak generation (e.g. using hydrogen), deploy bioenergy with CCS
Hydrogen	Start large-scale hydrogen production with CCS	Widespread deployment in industry, use in back-up electricity generation, heavier vehicles (e.g. HGVs, trains) and potentially heating on the coldest days
Buildings	Efficiency, heat networks, heat pumps (new-build, off-gas, hybrids)	Widespread electrification, expand heat networks, gas grids potentially switch to hydrogen
Road transport	Ramp up EV market, decisions on HGVs	Turn over fleets to zero-emission vehicles: cars & vans before HGVs
Industry	Initial CCS clusters, energy & resource efficiency	Further CCS, widespread use of hydrogen, some electrification
Land use	Afforestation, peatland restoration	
Agriculture	Healthier diets, reduced food waste, tree growing and low-carbon farming practices	
Aviation	Operational measures, new plane efficiency, constrained demand growth, limited sustainable biofuels	
Shipping	Operational measures, new ship fuel efficiency, use of ammonia	
Waste	Reduce waste, increase recycling rates, landfill ban for biodegradable waste	Limit emissions from combustion of non-bio wastes (e.g. deploy measures to reduce emissions from waste water)
F-gases	Move almost completely away from F-gases	
Greenhouse gas removals	Develop options & policy framework	Deployment of BECCS in various forms, demonstrate direct air capture of CO_2, other removals depending on progress
Infrastructure	Industrial CCS clusters, decisions on gas grid & HGV infrastructure, expand vehicle charging & electricty grids	Hydrogen supply for industry & potentially buildings, roll-out of infrastructure for hydrogen/electric HGVs, more CCS infrastructure, electricty network expansion
Co-benefits	Health benefits due to improved air quality, healthier diets and more walking & cycling. Clean growth and industrial opportunities	

Source: CCC analysis
Notes: CCS = carbon capture and storage.
EV = electric vehicle. BECCS = bioenergy with CCS

www.theccc.org.uk

Reforest an area the size of the US to help avert climate breakdown

THE CONVERSATION

***An article from* The Conversation**

By Mark Maslin, Professor of Earth System Science, UCL

Restoring the world's forests on an unprecedented scale is 'the best climate change solution available', according to a new study. The researchers claim that covering 900 million hectares of land – roughly the size of the continental US – with trees could store up to 205 billion tonnes of carbon, about two-thirds of the carbon that humans have already put into the atmosphere.

While the best solution to climate change remains leaving fossil fuels in the ground, we will still need to suck carbon dioxide (CO_2) out of the atmosphere this century if we are to keep global warming below 1.5°C. So the idea of reforesting much of the world isn't as far-fetched as it sounds.

Since the dawn of agriculture, humans have cut down three trillion trees – about half the trees on Earth. Already 43 countries have pledged to restore 292 million hectares of degraded land to forest worldwide. That's an area ten times the size of the UK. But what the new study advocates is reforesting something like ten times that amount.

Rewilding habitats and reforesting may be easier in the future as the world is already becoming a wilder place in many areas. This may seem a strange prediction, given that the global population will grow from 7.7 billion to 10 billion by 2050, but by then nearly 70% of us will live in cities and have abandoned rural areas, making them ripe for restoration. In Europe already, 2.2 million hectares of forest regrew per year between 2000–2015, and forest cover in Spain has increased from 8% of the country's territory in 1900 to 25% today.

Massive reforestation isn't a pipe dream and it can have real benefits for people. In the late 1990s, environmental deterioration in China became critical, with vast areas resembling the Dust Bowl of the American Midwest in the 1930s. Six bold programmes were introduced, targeting over 100 million hectares of land for reforestation.

Grain for Green is the largest and best known of these. It reduced soil erosion and stabilised local rainfall patterns. The ongoing programme has also helped to alleviate poverty by making payments directly to farmers who set aside their land for reforestation.

Better yet, the new study suggests that bringing back 900 million hectares of forest wouldn't impact on our capacity to reserve land for growing food. This is certainly possible, and in line with other estimates. Reforestation may even result in production from farmland increasing, as was found in China when more stable rainfall and fertile soil followed the return of forests.

No solution without emission cuts

There should be more scepticism about how much CO_2 900 million hectares of new forest could store though. The paper insists on 205 billion tonnes of carbon, but this seems too high when compared to previous studies or climate models. The authors have forgotten the carbon that's already stored in the vegetation and soil of degraded land that their new forests would replace. The amount of carbon that reforestation could lock up is the difference between the two.

Past and future carbon emissions

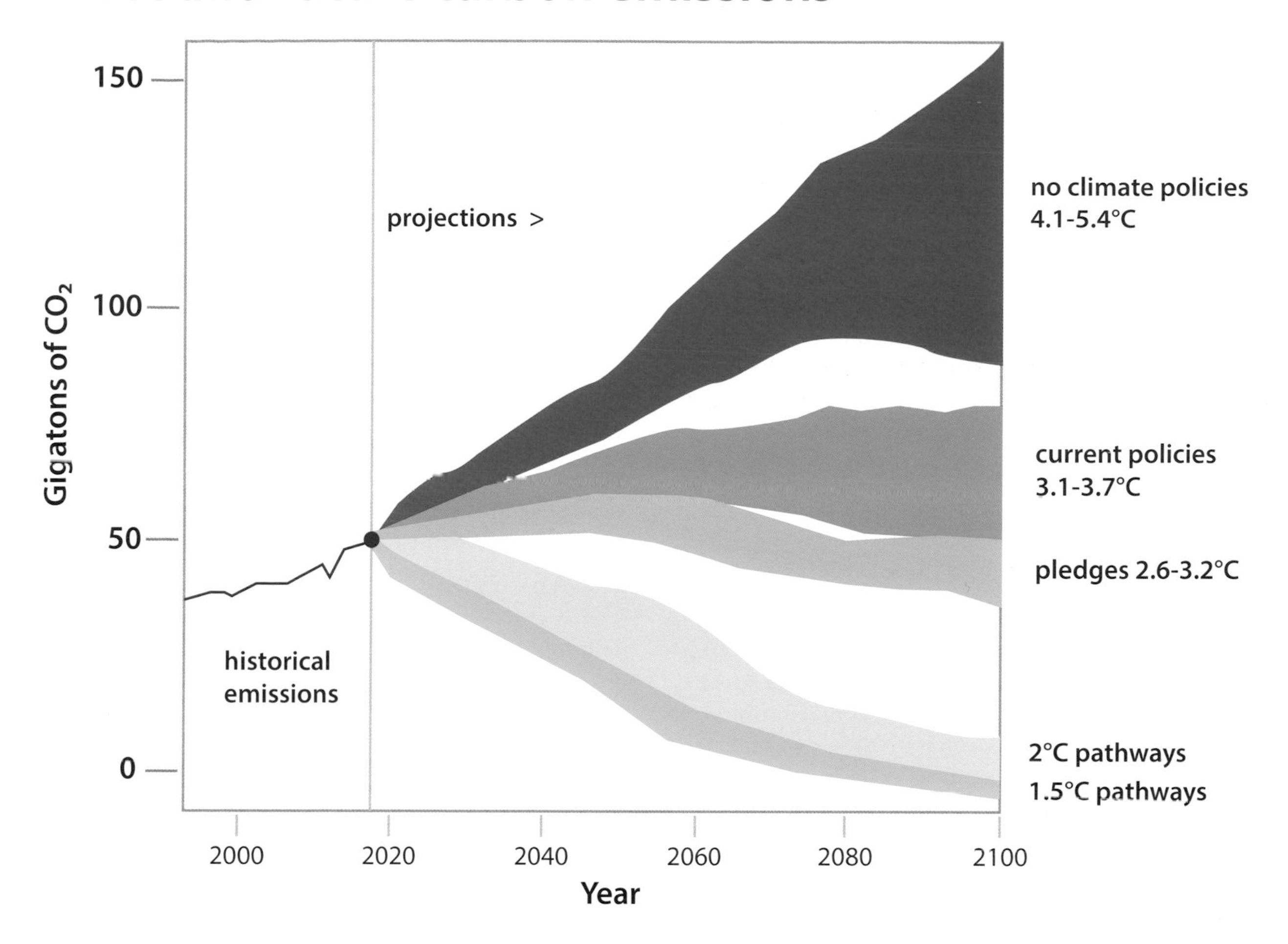

Mature forests can store a lot of carbon, but this capacity is only reached after hundreds of years, not a couple of decades of new forest growth as assumed in this study. The most recent estimate from the IPCC suggests that new forests could store on average an extra 57 billion tonnes of carbon by the end of the century. This is still a huge number and could absorb about one-sixth of the carbon that's already in the atmosphere, but reforestation should be thought of as one solution to climate change among many.

Radically reducing carbon emissions and absorbing the carbon that's already in the atmosphere will be necessary to avert catastrophic climate change.

Even if warming is stabilised at 1.5°C, the study indicates that one-fifth of the land proposed for reforestation could be rendered too hot for growing new forests by 2050. But this concern ignores the role of carbon dioxide fertilisation – when there are higher levels of carbon dioxide in the atmosphere, photosynthesis is more efficient, meaning plants need less water and can still be productive at higher temperatures. Today, the most immediate threat to tropical forests is deforestation by people and the fires they light which get out of control, not the more subtle impacts of higher temperatures.

Reforesting an area the size of the US will have massive benefits on local environments and will store a huge amount of man-made carbon emissions. It is not, however, a substitute for reducing those carbon emissions.

Even if the world reduces its carbon emissions to zero by 2050, there will still need to be negative global carbon emissions for the rest of the century – drawing CO_2 out of the atmosphere to stabilise global warming at 1.5°C. Reforestation is essential for creating negative emissions – not reducing the amount of carbon that humans are still emitting.

There is another sting in the tail. Massive reforestation only works if the world's current forest cover is maintained and increasing. Deforestation of the Amazon rainforest – the world's largest – has increased since Brazil's new far-right president, Jair Bolsonaro, came to power. Current estimates suggest areas of rainforest the size of a football pitch are being cleared every single minute.

It won't be easy, but society needs to protect the forests we've got, and protect new forests in perpetuity to permanently keep carbon sequestered in trees and out of the atmosphere.

4 July 2019

Key Facts

- The Climate Change Act was passed by the UK Government in 2008. (page 1)
- A recent IPCC report warned that humanity has 12 years to take emergency action in order to prevent global warming greater than 1.5°C. (page 2)
- Just two countries, China and the US, are responsible for more than 40% of the world's CO_2 emissions. (page 3)
- Before the Industrial Revolution, levels of atmospheric CO_2 were around 280 parts per million (ppm). As of 3 June 2019 it stands at 414.40ppm. (page 3)
- Qatar produces the highest per capita CO_2 emissions in the world. (page 4)
- The 2015 Paris Accord aims to keep global warming between 1.5°C and 2°C by the end of this century. But some experts are now warning that the 2°C limit is too lax and would leave swathes of the population lacking water and food from mid-century. A report from the UN Intergovernmental Panel on Climate Change (IPCC) said adopting the tougher target of 1.5°C is the only way to avoid that. (page 6)
- Findings from the European Social Survey (ESS) show that 93% of people acknowledge that the world's climate is at least probably changing. (page 7)
- 36% of ESS respondents say that climate change is mainly or entirely caused by human activity. A majority of people (53%) blame human activity and natural causes equally for climate change, with a vast majority (95%) thinking that climate change is at least in part caused by human activity. (page 7)
- Younger people (46%) are more likely than those over the age of 65 (27%) to think climate change is entirely or mainly due to human activity. (page 7)
- The UK already leads the world in tackling climate change – with emissions reduced by more than 40 per cent since 1990. (page 8)
- Since 2010, government has invested a record £2.6 billion in flood defences, and is on track to protect 300,000 more homes from flooding by 2021. (page 9)
- A hurricane is a large rotating storm that forms over tropical or subtropical waters in the Atlantic. (page 13)
- Storms are given names once they have sustained winds of more than 39mph. Hurricanes are gauged by something called the Saffir-Simpson hurricane wind scale, which runs from one to five and measures speed. (page 13)
- Category five storms, of at least 157mph, can raze dwellings, cause widespread power outages and result in scores of deaths. (page 13)
- Almost all hurricanes develop once the northern hemisphere approaches summer, with the hurricane season running from 1 June to 30 November. The season peaks between August and October. (page 13)
- The world lost 12 million hectares of tropical rainforest in 2018 – an area the size of North Korea and the equivalent of 30 football pitches every minute. (page 16)
- Intergovernmental Panel on Climate Change scientists estimated 99% of corals across the globe are likely to be lost if the climate crisis is not addressed and global heating reaches 2°C. (page 17)
- Globally, more than 100 million people live in coastal regions vulnerable to sea-level rise, and many of the world's largest cities are situated less than 10 metres above current sea level. (page 20)
- Together, the ice sheets in Greenland and Antarctica hold over 99% of all ice on Earth. (page 20)
- Cows and other ruminant animals (like goats and sheep) emit methane, a potent greenhouse gas, as they digest grasses and plants. This process is called 'enteric fermentation,' and it's the origin of cows' burps. (page 24)
- Beef is more resource-intensive to produce than most other kinds of meat, and animal-based foods overall are more resource-intensive than plant-based foods. (page 24)
- If ruminant meat consumption in high-consuming countries declined to about 50 calories a day or 1.5 burgers per person per week – about half of current US levels and 25% below current European levels, but still well above the national average for most countries – it would nearly eliminate the need for additional agricultural expansion (and associated deforestation), even in a world with 10 billion people. (page 26)
- Asphalt heats up 10–20°C higher than the air. (page 28)
- The 2015 heatwave in India killed more than 2,000 people. (page 29)
- One acre of wildflower meadow on a single day in summer can contain three million flowers, producing 1kg of nectar – enough to support nearly 96,000 honeybees per day. (page 33)

Glossary

Carbon footprint
A carbon footprint is a measure of an individual's effect on the environment, taking into account all greenhouse gases that have been emitted for heating, lighting, transport, etc. throughout that individual's average day.

Carbon offsets
Carbon offsets are a reduction in greenhouse gas emissions made in order to compensate for greenhouse gas production somewhere else. Offsets can be purchased in order to comply with caps, such as the Kyoto Protocol. For example, rich industrialised countries may purchase carbon offsets from a developing country in order to satisfy environmental legislation.

Carbon tax
Carbon tax is a fee imposed on the production, distribution and burning of fossil fuels responsible for CO_2 and other greenhouse gas emissions. It incentivises companies to cut their carbon emissions and invest in cleaner, greener options.

CO_2 emissions
Carbon dioxide gas released into the atmosphere. CO_2 is released when fossil fuels are burnt. An increase in CO_2 emissions due to human activity is arguably the main cause of global warming.

Climate change
Climate change describes a global change in the balance of energy absorbed and emitted into the atmosphere. This imbalance can be triggered by natural or human processes. It can cause either regional or global changes in weather averages and frequency of severe climatic events.

Climate change refugees
Also known as 'environmental migrants', climate change refugees are people who have been forced to flee their home region following severe changes in their local environment as a result of global warming.

Climate models
Scientific models which are designed to replicate the Earth's climate. Scientists are able to hypothetically test the effects of global warming by simulating changes to the Earth's atmosphere.

Extinction Rebellion
Extinction Rebellion (abbreviated as XR) is a global movement of climate activists using civil disobedience and non-violent resistance to protest against climate breakdown and prevent mass extinction.

Geoengineering
Geoengineering means large-scale interventions in the Earth's climate system to try to tackle climate change. There are, broadly, two types. The first would try to take some of the greenhouse gases that are causing climate change out of the atmosphere. The second would try to balance out that we have too many greenhouse gases in our atmosphere by reflecting more sunlight back into space.

Global warming
This refers to a rise in global average temperatures, caused by higher levels of greenhouse gases entering the atmosphere. Global warming is affecting the Earth in a number of ways, including melting the polar ice caps, which in turn is leading to rising sea levels.

Greenhouse gases (GHG)
A greenhouse gas is a type of gas that can absorb and emit longwave radiation within the atmosphere: for example, carbon dioxide, methane and nitrous oxide. Human activity is increasing the level of greenhouse gases in the atmosphere, causing the warming of the Earth. This is known as the greenhouse effect.

IPCC
An abbreviation for Intergovernmental Panel on Climate Change, the leading scientific body which assesses and reviews global climate change. It was founded by the United Nations Environment Programme and the World Meteorological Organization and currently has 194 member countries from around the world.

Net zero
'Net zero' is the target many countries have set themselves to achieve by 2050. It refers to achieving an overall balance between emissions produced and emissions removed from the atmosphere.

Paris Agreement
The Paris Agreement, also referred to as the 'Paris climate accord' is a pact, established in 2015 and sponsored by the United Nations, to bring the world's countries together in the fight against climate change.

Assignments

Brainstorming

In small groups, discuss what you know about the climate emergency. Consider the following points:

- What is climate change?
- What is the 'greenhouse effect'?
- Who is most responsible for causing climate change?
- What is the international community doing to combat climate change?
- What is 'Net zero'?

- Create a mindmap showing all that you know about the causes and consequences of climate change.

Research

- Design and present a ten-minute PowerPoint presentation on the predicted future impacts of climate change. Explain what is expected to happen to global temperatures and sea levels.
- What is your carbon footprint? Visit www.carbonfootprint.com to find out your carbon footprint. How could you reduce this? Compare your findings with others in your class. Is there much variation throughout your class or do you all have similar footprints?
- Not everyone believes there is a climate emergency. Who are the most prominent climate change deniers in the world today? Can you find out the reasons why they reject the notion of a global climate crisis?
- Do some research about extreme weather events, for example: storms, floods and heatwaves. Create a timeline recording all of the major, most destructive events that have occurred around the world so far this century.

Design

- Design a wall poster giving people information about ways to combat climate change in their daily lives. Include suggestions about energy saving and recycling. You should make it readable and attractive, suitable for a teenage audience. You can add graphics, illustrations or diagrams to your leaflet to make it more accessible.
- Read the article on page 28: *Halfway to boiling: the city at 50°C* and create an illustration to highlight the extreme effects of rising temperatures as described in the piece.

Oral

- In small groups, role play a radio talk show on the topic of climate change. One student will play the radio show host, another a climate change expert who aims to dispel some common myths, and a third student will be a climate change sceptic arguing that human actions do not cause global warming. Other students can play listeners phoning in with questions. The host should aim to stimulate a lively debate on the topic, giving equal time to all arguments.
- As a class, make a list of small changes that can be made on a day-to-day basis to combat climate change. Discuss which actions they believe are the most effective and if they can identify areas where their school or college could do more to improve their 'green' credentials.
- 'Children should be allowed time off school to join the climate strike' Debate this motion as a class, with one group arguing in favour and the other against.

Reading/writing

- Watch Leonardo DiCaprio's 2016 climate change documentary *Before the Flood*. Do you think this film is an effective way of conveying the climate change message? Write a short paragraph reviewing this film.
- Write a short essay on each or all of the following
  - Climate change and the destruction of the rainforest
  - Climate change and the disappearing Great Barrier Reef
  - The effects of climate change on plants and animals
  - Climate change and the effect on human health
- Thinking about the research you have done and the articles you have read, write a list of the things you now know about climate change that you didn't before.

Index

Acknowledgements

The publisher is grateful for permission to reproduce the material in this book. While every care has been taken to trace and acknowledge copyright, the publisher tenders its apology for any accidental infringement or where copyright has proved untraceable. The publisher would be pleased to come to a suitable arrangement in any such case with the rightful owner.

Images

Cover image courtesy of iStock. All other images courtesy of Pixabay, rawpixel.com and Unsplash.

Illustrations

Don Hatcher: pages 24 & 27 Simon Kneebone: pages 4 & 10 Angelo Madrid: pages 1 & 9.

Additional acknowledgements

With thanks to the Independence team: Shelley Baldry, Danielle Lobban, Jackie Staines and Jan Sunderland.

Tracy Biram

Cambridge, September 2019